Arab Paper

Arab Paper

Joseph von Karabacek

Translated by
Don Baker and Suzy Dittmar

Additional Notes by
Don Baker

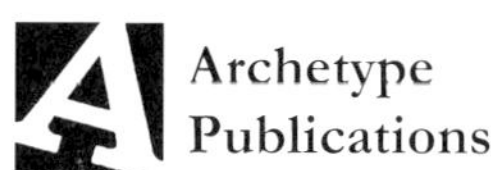

First published (in wirobound format) by Islington Books Ltd, 1991
This edition published by Archetype Publications in association with the Don Baker Memorial Fund 2001

Archetype Publications Ltd.
6 Fitzroy Square
London W1T 5HJ

www.archetype.co.uk

Tel: 44(207) 380 0800
Fax: 44(207) 380 0500

Archetype Publications is grateful to the Don Baker Memorial Fund for a donation towards the costs of publication of this volume.

ISBN 1-873132-02-6

British Library Cataloguing in Publication Data
A catalogue record for this book is available from the British Library.

Typeset by Kate Williams, Abergavenny
Printed and bound in Great Britain by Antony Rowe Ltd., Chippenham

Contents

Acknowledgements

This article on Arab paper was first published in 1887 as *Das Arabische Papier* in *Mitteilungen aus der Sammlung der Papyrus Erzherzog Rainer* 2/3, Vienna.

May I give my thanks to the many people who have helped with this translation. To Dr Richard Fisher for his continual encouragement; to Yasin Safadi for frequent technical advice. Particular thanks to Dr Adam Gacak and Dr G. T. Scanlon for reading the proofs and to Dr D. Priest for comments on paper history.

Preface

Bibliophiles in Europe at the end of the nineteenth century began to take a serious interest in the manuscripts of the Middle East and the paper on which they were written. Perhaps the most important of these men were C.-M. Briquet working in Geneva, and J. Wiesner and J. von Karabacek working in Vienna. All three were concerned with the burning topic of the moment: was oriental paper made of cotton? Within the space of two years these three writers published the seminal articles for the European study of Arab paper. *Das Arabische Papier* was one of those articles.

My reason for starting this project was simply the desire to know the contents of this much quoted article. Suzy Dittmar supplied a basic translation from the German. My task was to remove some of the more convoluted expressions in which J. von Karabacek excels and to attempt to provide a more modern and accessible translation of the Arabic words and names.

Don Baker
London 1991

Publisher's note

There are two sets of notes in this volume. The original author's notes are indicated in the text by Arabic numerals while the translator's notes are in Roman numerals (in brackets).

1

Introduction: The Rainer Collection

The Rainer Collection contains many treasures in manuscript and a remarkable variety of paper. The genius of the Arabs carried the art of papermaking from the borders of the Heavenly Kingdom across to the West. The collection, unique in its unity and importance, contains an enormous number of documents on paper from the start of papermaking up to the late Middle Ages, showing the development of this writing material over a period of 600 years. I have sorted and studied 12,000 papers from the Rainer Collection, and there are still more. The earliest pieces can be dated to the 2nd century after the Hijra (796–815AH) and the latest to the year 1388AH – it is a continuous chain with no missing links. It was inevitable that such an unrivalled collection should be examined by scientific methods.

Western researchers of paper history had found little to help them from Western historical sources, and only the scientific study of the original material could be expected to produce results. Enlightenment has finally arrived – and from a surprising direction. Following a three-year study, Professor Wiesner made a public statement about his discoveries.[1]

What Wiesner saw in those remnants of paper was revealing in many ways and historical researchers have reason to be grateful to him. The foundations he laid by scientific means will be built upon. His results have proved that many of the basic ideas held about the history of papermaking were wrong. The criteria used for paper classification were equally wrong. He gets rid of old errors and presents new evidence which will eventually be accepted by even his greatest critics. Wiesner destroys the old misconceptions about the existence of cotton paper, and this alone makes his scientific work of lasting importance.

It is a pleasure to see scientific criteria being applied for the first, but probably not the last, time in the examination of historical documents. Cooperation between the two different methods of research – scientific and historical – becomes imperative when we leave Western sources, overworked and dubious as they are, and turn instead to the origin of our paper, the Orient itself. Here we enter an area about which paleographers know little, and that only hearsay. In my opinion, we underestimate the importance of Oriental studies for enlightenment on various cultural and historical problems of concern to us. Yet not so long ago, universities considered excluding Oriental studies from their curricula.

In the course of the following pages, I will try to show that the Oriental documents that we have can supply answers where Western sources can be of no further assistance.

Examination of this fortuitous discovery of old documents allows us to consider the early days of the Arab use of paper and to fix a date for the victory of this new writing material over papyrus and parchment.

Most of the papyrus in the Rainer Collection came from Arsinoë-Fayyūm, but only a very few of the papers come from there. In a previous article (*Mitteilungen* I, 66, 107) I have described where most of the paper was found – the district of Ashmūnain and its main town.

Ashmūn, or popularly Ashmūnain, was the heavily populated capital of a province of Middle Egypt.[2] An important centre of manufacturing, it was situated three staging points away from the west bank of the Nile, near the present St. Joseph's Canal.[3] Today the town has no more that 10,000 inhabitants and belongs to the administrative district of al-Minyah.

The paper was found buried in the ground at Ashmūn, according to the Arabs who discovered it, and this is confirmed by the documents themselves, the town being mentioned in a variety of references. There are tax demands for a particular year coming 'from the Diwān of Ashmūnain', or referring to 'both administrative districts of Ashmūnain'. A tax payer is charged 'according to the tax he is obliged to pay in al-Ashmūnain'; later he received a receipt. Often the accounts are dealt with in 'the administrative district of Ashmūnain', which is ruled by the 'governor of Ashmūnain'. There are summonses to 'the Law Court of the town of Ashmūnain' which give evidence of legal procedures of the town. People calling themselves 'born Ashmūnainis' send and receive letters, or appear in administrative documents. The name of the town appears on letters to and from Ashmūnain: for example, a slave writes 'by the Grand Mosque of Ashmūnain'. A second lady, 'the mother of Husain', writes a letter to that same son living in the Faraj Mosque quarter of the town. Another comes from 'the big market place in the mosque quarter' and many other documents bear the name of the Ashmūn province. The commercial papers mention the area too; for example, the bills issued in al-Ashmūnain in which '250 containers of grain for al-Fayyūm' are mentioned. The astrologer 'Umar ibn Sulayman calculates in the Ashmūnain pound-weight system in a paper dating to the 9th century. This evidence plus much more shows clearly the origin of the papers.

Not only the contents but also the form, colour and manner of writing lead to the attribution of these pieces to the area of al-Ashmūnain, mainly its capital but also various other district offices, Ibshāda, Dalga, Tahā, Darūt, Ansina, Antinoë etc.

Obviously it is important to be absolutely sure of the provenance of the papers when classifying the Coptic material. The latter forms only a small part of the collection, currently 500, compared to 12,000 Arabic pieces.

It would take too long to describe the contents of these papers. Most of the pieces are not well preserved but even the smallest pieces of *disjecta membra* are unique and capable of rewarding the reader with something of interest. There are, however, hundreds of documents in beautiful condition.

Regardless of whether the papers are complete or fragmentary, they revolutionise the study of Oriental documents. Through them, questions of great cultural and historical importance, such as the invention of printing, can be examined in a new light. The great age of some pieces in the Rainer Collection adds to their importance. These ancient manuscripts come almost from the very beginning of Arab papermaking and are followed by a chronological list of dated papers similar to the one I was able to produce for papyrus (*Mitteilungen* I). This list makes an important contribution to establishing the date at which papyrus manufacture ended in Egypt. These papers present a unique research opportunity, the history of the writing material in tangible form.

The earliest pieces in the collection are well preserved but undated. However, the experience gained in the classification of the dated papyri helps us because the handwriting of a period always shows certain constant characteristics.

The oldest paper I have found so far is undated, but its characteristics point to some time between 180H and 200H (795–815AH). One or two others also belong to the 2nd century Hijra, and about two dozen have been confirmed as coming from the 3rd century H (9th century AH). The earliest possible date for Paper 7800 is assumed to be 203H (819AH); this is not absolutely certain due to an ambiguity in certain passages arising from the use of Greek number letters ϲγ سه that is, ϲγ تبه meaning '*Tybi* (January) 203'. The figures 203 might not refer to the year but be a note of something paid or received each month. The reading سنه , year, for تبه is not possible.

Paper 4092 mentions Sulaimān ibn al-Ash'adh al-Sijistānī, one of the six Traditionists, born 202H (817AH), died 275H (889AH). He visited Egypt on his travels collecting the Hadith, accompanied and assisted by his son Abdullah, born 230H (845AH). This dates the paper more or less certainly. Other 3rd century pieces are dated *Rajab* 287 (between 2 and 31 July 900AH), and another bears the date *Safar* 296 (November 908AH).[4]

The papers of the 4th and following centuries are easier to classify because they are frequently dated. These begin, as far as I can ascertain, in 300H (912AH) and end 790H (1388AH). The collection is particularly strong in 4th century material and provides the links needed to reach conclusions about the period of any of the undated papers.

The papers are dated in two ways, in Arabic words or in Greek number letters:

τλ – 307H = 919AH
τλζ – 326H = 938AH
τλ – 330H = 942AH
τλζ – 337H = 948/9AH
τλη – 338H = 949/50AH
τλθ – 339H = 950/1AH
τμβ – 342H = 953AH
τμδ – 344H = 955AH
τμη – 348H = 959/60AH
τμθ – 349H = 960AH
τν – 350H = 961AH
τνϛ – 356H = 967AH
τνη – 358H = 969AH
τϙ – 390H = 1000AH
τκζ – 427H = 1036AH
φιγ – 513H = 1119AH

On first impression, the date φιγ appears to be late, but there is evidence that the use of Greek number letters had become widespread because they were employed in government offices in Egypt. They can be found in the Arabic papers of the Rainer Collection through to the 14th century AH. Because we are currently unfamiliar with Greek documents of the Middle Ages, the importance of these accounts cannot be underestimated. They provide information on Greek cursive scripts as opposed to the more familiar book scripts.

Of course, dating is usually in the Arabic form, the year transcribed into words, and when this is used, the chronology can be established with relative certainty. The order of numerals in dates always begin with the units, for example 'five and twenty and three hundred' whereas the opposite order, 'three hundred and twenty five' is usually used for accounts. I stress this because the recognition of the date is one of the most difficult tasks of the translator. In the course of twenty years in the study of paleography, I have never seen such difficult texts as these Ashmūn documents. From the 4th century onwards the documents are written usually in the so-called *Qarmati* script which is wrongly assumed to be the 'formal epigraphic' or 'ornamental Kufic' of epigraphic inscriptions. The difficulties in reading the script are: the loss of distinction between individual letters, the ligatures caused by not lifting the pen, the backward strokes and the frequent use of *Comes largitionum sacrarum*. The combining of letters is added to this producing an almost impenetrable riddle for the expert in Arabic scripts and greatly tests the deductive capacity of the translator.

In short the script of these documents is an intertwined and compact chancery, or Dīwānī hand. *Qarmati* is just the older name for the cursive, Dīwānī script to which it is identical in all respects.[5] Dīwānī is wrongly seen as a Turkish invention for use in diplomacy.

With reference to the chronology of the papers, it might be useful to show the following papyrus dates. There is always more than one example for any year.

Papyrus:

Indiction (i) year XV	(641–642AH = 20–21H)
Indiction year I	(642–643AH = 21–22H)
Indiction year II	(643–644AH = 22–23H)
Indiction year VI	(647–648AH = 27–28H)
Indiction year XI	(652–653AH = 32–33H)
Hijra years:	22, 30?, 73, 84, 90, 91, 95

2nd AH 104, 106, 111, 116, 117, 119, 124, 125, 130, 137, 142, 144, 150, 154, 162, 164, 169, 170, 171, 174, 175, 176, 177, 178, 179, 180, 183, 189, 191, 192, 194, 196, 197, 198, 199

3rd AH 200, 201, 202, 203, 204, 205, 206, 207, 208, 209, 210, 211, 212, 213, 214, 215, 216, 217, 218, 219, 220, 221, 222, 223, 224, 225, 226, 227, 228, 229, 230, 231, 232, 233, 235, 236, 237, 238, 239, 240, 241, 242, 243, 244, 245, 246, 247, 248, 249, 250, 251, 252, 253, 254, 255, 256, 257, 258, 259, 260, 261, 262, 263, 264, 265, 266, 267, 268, 269, 270, 272, 273, 274, 275, 276, 277, 278, 279, 280, 281, 282, 283, 284, 285, 286, 287, 288, 289, 290, 291, 292, 294, 297, 299

4th AH 305, 306, 308, 310, 311, 314, 319, 320, 323

Paper:

4th AH 300, 302, 303, 304, 305, 306, 307, 315, 320, 323, 324, 325, 326, 327, 328, 329, 330, 331, 332, 333, 335, 336, 337, 338, 339, 340, 341, 342, 343, 344, 345, 346, 347, 348, 349, 350, 351, 352, 353, 354, 355, 356, 357, 358, 359, 361, 362, 363, 364, 365, 366, 367, 368, 369, 371, 373, 374, 375, 376, 377, 381, 382, 384, 385, 386, 388, 389, 390, 394, 395, 396, 397, 398, 399

5th AH 400, 401, 402, 403, 404, 405, 406, 407, 408, 409, 410, 411, 412, 413, 414, 415, 416, 417, 418, 419, 421, 422, 423, 424, 425, 426, 427, 428, 430, 431, 432, 433, 434, 435, 436, 437, 438, 439, 440, 441, 442, 443, 444, 445, 446, 447, 449, 451, 460, 464, 470, 475, 480, 482, 490

6th AH	500, 504, 506, 507, 513, 517, 521, 530, 540, 543, 545, 561, 584, 592
7th AH	600, 603, 604, 608, 610, 612, 613, 616, 619, 621, 622, 624, 626, 627, 634, 635, 637, 639, 640, 641, 642, 643, 644, 649, 651, 657, 660, 661, 664, 665, 666, 668, 673, 674, 678, 679, 680, 681, 683, 685, 686, 687, 688, 689, 690, 691, 692, 693, 695, 696, 698
8th AH	700, 703, 704, 705, 706, 707, 711, 712, 713, 714, 715, 716, 724, 725, 726, 727, 728, 730, 732, 734, 736, 740, 741, 742, 743, 744, 745, 747, 750, 751, 757, 764, 769, 771, 780, 782, 783, 785, 789, 790

We must acknowledge that this collection of documents stretching over 800 years gives a great insight into records during the period of brilliance in the Land of the Nile. Not only do they present a vivid picture of the scale of Arab papyrus production in Egypt during post-Byzantine time (as I stressed in *Mitteilungen*, I, 50), but they are also useful in settling the much discussed question of the duration of papyrus manufacture. These dates show the time limits for the continued dominance of papyrus in the western regions of the Caliphate and the Western world despite the arrival of paper in the eastern provinces. The figures show both materials competing until paper, a cheap and technically superior writing material, won a victory over papyrus, finally replacing an indigenous writing material that had been used in Egypt for thousands of years. At present we have settled on the date 323H (935AH); this is the last year for which we have dated papyrus. Paper had appeared, as witnessed by our collection, at the beginning of the 4th century, and its challenge to papyrus increased year by year. It had probably reached Egypt as a new but not very successful item of trade at the end of the 2nd century and through the 3rd century AH. We can therefore assume with some degree of assurance that Egyptian papyrus-making was virtually ended by the middle (the second half) of the 10th century AD.

Now let us consult our Arabic sources on this matter. We learn that paper had not managed to replace papyrus in the 3rd century AH (9th century AD) because papyrus was still being made in the first half of that century (as our documents prove) and being made without competition from paper, according to al-Kindī (d. 26H, 860AD). The size of the rolls of papyrus had not even decreased by then; the rolls were made with a length of 30 ells (14.5m.) or larger, with a width of a span (measured between the thumb and the index figure).[6]

Caliph al-Mu'tasim bi-Allah decided, in the year 221H (836AD) to leave his old residence by the Tigris and found Sāmarrā about three days' journey to the north of Baghdād.[7] Artisans and traders were moved as well to add the necessary lustre to this town of palaces. Al-Ya'qūbī, writing in 891AH, mentions this colonisation.[8] An attempt was made, with the help of Egypt, to introduce papyrus manufacture to the new residence, either by importing the reed from Egypt (compare the papyrus factory in Rome during Pliny's time), or by growing the reed on the Euphrates or in Palestine.[9] This is seen as evidence that the papyrus industry withstood the large quantities of paper imported from the east, and that long-established habit could not be easily changed by a new invention.

Two different cultures were created, each writing material claiming half the world. This is made clear by a quotation from al-Jāhiz (d. 255H, 869AD): 'Sheets of Egyptian papyrus are to the West what Samarqand paper is to the East'.[10] In this way, he names the centres of the two writing materials. Samarqand is the capital of the land of paper, and it will be shown that from here, paper started its journey around the world. Ibn al-Faqīh writes as late as 903AD in his book on the geography of Egypt: 'They have papyrus sheets and no one can compete with them'.[11] Egypt was still providing the nearer eastern provinces of the Caliphate with papyrus while Hārūn al-Rashīd and his officials were using paper for their work. Papyrus vendors were still doing business in al-Karkh, the well-known merchants' area of Baghdād, during the years 816–17AD.[12] Abū Nasr Bishr ibn al-Hārith (d. 227H, 842AD)[13] recounts how he picked up a papyrus sheet from the ground in Baghdād, washed it in rose water and perfumed it with musk.[14] Throughout the 9th century both writing materials were in equal use; in addition to this, there was parchment. It was almost the beginning of the 10th century when the Isfahāni, 'Alī ibn al-Azhar (d. 307H, 919–20AD) described to a friend the various kinds of pens and their use on paper, parchment and papyrus.[15]

A similar situation existed in the western areas of the Arab world, even in Spain. There was strong competition between the two writing materials, and it should be noted that by the beginning of the 10th century, paper as well as papyrus was well known and in use at Guadalquivir.[16] For example, the delightful book, *al-'Iqd*, written in Spain, which is, as its title suggests, a necklace or anthology of Arabic literature. Its author Ibn 'Abd Rabbihī (860–940), born in Cordova, wrote not only that letter on the various kinds of pens for paper and papyrus, but also provides a great deal of expert information on that other contemporary writing material, papyrus.[17] Papyrus could not compete with paper after the beginning of the 10th century. This was the start of its final decline which we can follow clearly in our collection of Arabic documents.

2

The decline of papyrus and the development of paper

The development of paper ran parallel to the rapid decline in the quality of papyrus. The collection shows that papyrus structure had quite degenerated: nothing is left of its former fineness, thinness and evenness. The overlapping strips are thick, frequently spongy, carelessly sorted, unevenly laid and imperfectly adhered and smoothed. There are more rough gloss-less sheets than fine.[18] Any fine ones are always palimpsest pages from a much earlier date that have already been used on one side and are now written on again. It is no coincidence that the last Arabic papyri are almost always palimpsests.

This deterioration in quality together with our chronological studies show that time had run out for 'Nile Paper'. Arab sources confirm that significant changes took place in the middle (the second half) of the 10th century. The much-travelled, well-educated and communicative al-Muqaddasī does not mention papyrus among the products exported from Egypt while dealing with Egyptian trade in his delightful book written in 375H (985/6AD).[19] Ibn al-Faqīh, however, mentioned it in 903AD. In the later source, the excellent pens were listed among 42 Egyptian specialities, but these did not include papyrus. In another place however, al-Muqaddasī mentions paper with some enthusiasm.

It does not prove anything when the author of the *Fihrist*, written two years later in 377H (987AD), said that 'the Egyptians wrote, and still write, on Egyptian papyrus made from papyrus reed'.[20] Certainly, its manufacture had not finished completely by then. For some time to come, it was used medically and for talismans, but its importance as a writing material and its export had decreased considerably.[21]

It is true, therefore, what Ibn Hawqal wrote in 367H (977/8AD) on the Sicilian papyrus reed for the geography book he edited: he knew nothing in the whole world which compared to the Egyptian papyrus plant, with the exception of the Sicilian reed.[22] Here he is talking about the plant, not the writing material derived from it. He does not even mention papyrus in the chapter on Egypt and its products written after he had visited the country in 359H (969AD). His choice of words makes it clear that he saw the plants that were cultivated in Egypt simply as a reminder of a glorious past.[23]

One simply cannot ascribe the decline of this craft to the 10th century, in accordance with the statement made by Eustathios at the end of the 12th century: 'their skill has now been forgotten'.[24] Surely these words mean nothing more than we learn from his contemporary, al-Nabbātī, a botanist who went to Egypt in 613H (1216AD); that is, that papyrus manufacture had stopped even though the plant itself still grew.[25]

I agree with Wattenbach as opposed to Delisle and Sickel that the *carta tomi* granted to the Corbie monastery by the Merovingian king, Childeric II, in 716 was on papyrus, that is to say, rolls of papyrus equivalent in size to the Arab *tumus*, a portion of a *qartās*.[26] The word *qartās* in our documents refers to the complete roll as it comes from the manufacturer. Wattenbach knew no other references to papyrus on this side of the Alps with the exception of a quotation dated 862 about which he has his doubts. It concerns Germany, a detail which made Sickel suspicious and led him to believe the quotation was of no significance. It deals with the words which, at the meeting of kings Ludwig and Lothar at Mainz in 862, the bishops added to the letter from the kings to the Pope. They wrote that they had been in a hurry: *unde etiam actum est, quod non iuxta morem antiquum in tuncardo conscripta cernitur (epistola) sed in membranis.* The otherwise unknown word *tuncardo* can only mean papyrus in this context, according to Wattenbach; this would mean that the use of papyrus would have been required by etiquette for correspondence with the Pope. Again, I agree with Wattenbach's interpretation of these words. In my opinion, *tuncardo* is only a variation of *tumario*, in Arabic *tūmūs*, and denotes an even smaller piece of papyrus. The term is often used in our papyri. The reference might be to 'a *tūmār* of the papyrus roll' or 'one *tūmār* from the papyrus roll'. Thus the Egyptian Director of Finance, al-Hasan ibn Sa'īd, orders in the 1st of Muharram 196H (23rd September 811AD):

> In the Name of God, the Compassionate, the Merciful
>
> Give my messenger one third of a *tumar* of clean Papyrus roll, God willing
>
> Written 1st Muharram 196
> bearing the seal: al-Hasan ibn Sa'id
> who believes in God and His Messenger

Twelve similar orders dating from the years 811 to 815, and some other documents allow us to calculate the divisions of a roll, or *tūmār*, of papyrus and to give the current prices for it.

Roll	Division		Price in		
qartās	tumus	tomario	dinar	carat	dirham
1	–	–	1/4	6	3
2/3	1	–	1/6	4	2
1/2	–	–	1/8	3	1 1/2
1/3	1/2	–	1/12	2	1
1/6	–	1	1/24	1	1/2
1/12	–	1/2	1/48	1/2	1/4
1/18	–	1/3	–	1/3	1/6

We see from this table that in Egypt during the first half of the 9th century, a roll of papyrus cost 6 gold carats, or quarter of a dinar at 4.25 normal grams, that is 1.0625 gold grams. A *tūmār* was about 2.42 metres long and cost about 54 centimes; 1/3 *tūmār* was 80.5cm long and cost about 18 centimes. The export of rolls of papyrus was therefore extremely lucrative for the finances of Egypt. During the course of the 9th century, it achieved wide distribution in the West: the last Frankish papyrus documents are from the year 862; a letter from Pope Nicolas 1 written on 28th April 863, and a papal bull from Stephen VI from the year 891 are also written on papyrus.[27] Papyrus was used exclusively for papal bulls as late as the 10th century, as Wattenbach has shown for the years 972 and 973. This writing material was already in short supply by the beginning of the 11th century, and parchment was being used. The use of papyrus had not completely stopped by the time of Victor II, 1055–57. It seems more than doubtful that the papyrus used at such a late date by the Papal Office can be of Egyptian manufacture. From my earlier studies it seems to me that it was more likely to have been of Sicilian origin.

We have reason to believe this because of evidence from the second half of the 10th century. It gives us important information about the cultivation of the papyrus plant at Palermo and about the manufacture of writing material from it. This is to be found in the reference I mentioned earlier by Ibn Hawqal:

> and the marshes of Palermo are completely over grown by this 'papyrus' (*bardīr*), that is the bardīr reed from which the *tūmār*, or sheets are made. I know of nothing in the whole world which compares to the Egyptian papyrus plant with the exception of the Sicilian reed. Much of it is used for ships' ropes, but a very small amount is made into tūmār for the sultan, depending on his needs.

These words by Ibn Hawqal do not contradict my hypothesis since he went to Palermo between 972 and 973 and the Sultan did not allow the export of papyrus, as happened later and continued up to the 13th century.[28]

It is different for the previously mentioned bulls dated 972 and 975. According to a conscientious and observant traveller, the papyri cannot have been of Sicilian origin. Also, they are of such a late date that they could not have been of Egyptian manufacture. Therefore, we can only assume that supplies of Egyptian papyrus rolls were no longer available and they were using the remnants of old supplies; it is well known that papyrus remains usable for a long time. Many pieces in the Rainer Collection still have usable surfaces and remain very supple even after a thousand years – all this despite poor conditions for their conservation. From the start of the decline of the Egyptian papyrus industry, the West was forced to use papyrus from supplies that were decades old. There is a striking example of this from the Papal Office: the famous bull of John VIII from the year 876, 3.9m long, given to the monastery of Tournus, and now in the Bibliothèque Nationale, Paris. There is a large piece of a factory mark in Arabic at the top giving the place of origin and its authentication, the signature guaranteeing the authenticity of the maker. A Muslim religious formula is also included, which seems not to have bothered the leader of Western Christianity.

This document has been examined many times. So far Amari's attempt at interpreting what is written has been the most useful for the classification of its writing material. Even so he could only decipher the words 'Allāh' and 'Sa'īd ibn'.[29] I realised at once that it was of Egyptian origin when I saw the illustration given by Champollion-Figeac.[30] The large collection of manuscripts found at al-Fayyūm has provided us with many similar pieces of evidence from Byzantine and Arab times. The bull fits exactly in size and character into the line of 9th century papyri. The width of the roll is the same: according to Amari, 58cm, or 2 old Parisian feet, that is 65cm according to Champollion-Figeac. A manuscript of one-third of a sheet width, that is 20cm, in the Rainer Collection points to a total width of 60cm. Amari was right (*loc. cit.*): the double red lines set between the lines of text and the strong, spirited hand, rich in ligatures corresponds to the Arabic script of that period. The words of the partially preserved first line also confirm the classification. It reads as follows: 'by order of Sa'īd ibn 'Abd al-Rahmān'.

The Arabic formula 'by order' implies here, as in all the documents where it is used, 'the Director of Finance in Egypt'. He was the successor of the *Comes largitionum sacrarum* of Byzantine times, supervising all the papyrus workshops of an area authorised by the Governor. Just as with the *Comes largitionum*, he was entitled to add his name to the manufacturer's marks on the rolls.[31] Although the date is missing due to the bad state of preservation, fortunately in this case we have been able to find out when Sa'īd ibn 'Abd al-Rahmān was in office. A beautiful document in the Rainer Collection from the year 223H (838AD) bears a seal.

Line 4: To Ishāq son of Simeon

5. in the presence of Sālih ibn al-Walīd who is the representative of 'Abdullāh
6. ibn Khalaf, Tax Collector for Sa'id ibn Abd al-Rahmān who was freed by the Amir of the Believers[32]

This document shows without doubt that Sa'id ibn 'Abd al-Rahman named on the manufacturer's mark on the bull of John VIII must be the same person as the Director of Finance of the same name. In consequence, it follows that the papal bull must have been written on a roll which is much older than its date, 876AD, leads us to believe. There is an interval of 38 years between this date and the year of office of Sa'id, 838AD, as stated on Papyrus 4965 – perhaps a few years more rather than less, because in 839AD another Director of Finance was nominated.

The Egyptian origin of the roll is therefore proven and the theories of Amari and of Pauli concerning its Sicilian origin are unfounded. The hypothesis of the latter concerning a papyrus industry on the island dating back to the 8th century is equally unfounded.[33]

Looking critically at the small amount of evidence we have, there is no possibility of such an industry in Sicily. Ibn Hawqal remarks that papyrus was cultivated for the personal use of the Sultan as opposed to the wider maritime use of the bardī reed (papyrus plant) in Palermo. This should have made it clear to us that conditions were not right for the long-term development of papyrus manufacture in Sicily, certainly not in the last third of the 10th century. From another point of view, it could not even have started at the beginning of that century because that ever-critical observer, Ibn al-Faqīh, states positively in 903, as we have seen before, that Egyptian papyrus manufacture was free of all competition.

The start of Sicilian papyrus production seems without doubt to be connected to the decline of the industry in Egypt.[34] Perhaps this decline was followed by an exodus of papyrus workers taking the bardī reed with them to Sicily in an attempt to obtain a living abroad using the skills of their forebears. This action would not be without precedent in the history of the Oriental crafts.[35]

I have been unable to find any enlightening references to the standing of paper compared to local Sicilian papyrus. It would be too much to expect that during its progress across North Africa to Spain, paper would not have entered Arab Sicily.

This leads us to the following question: how much time divides the arrival of paper among the Arabs from its final victory over papyrus? When I tackle the question of its first appearance and its eventual spread throughout Islam, I am aware that I am taking on one of the most difficult and shadowy chapters in the cultural history of mankind. Much has been written on the subject already. I have tried through effort of intellect to produce more than just an hypothesis while endeavouring to lift that thick veil that covers the history of paper. Fierce scholarly discussion on the content of the oldest paper was as unyielding as was investigation under the microscope or examination by the sense of touch.

Wiesner has given a summary of the historical study of old paper in his first chapter, so I have no need to repeat it.[36] Let me say, however, according to established opinion, there were two fundamental statements, the most important in the whole of paper history, which were generally considered to be proven. They were:

1 that the oldest matted papers were made of cotton and
2 that cotton papers were the antecedents of 'rag' paper, the invention of which can be ascribed to the Germans or the Italians.

Let us consider the earliest appearance of paper in central Asia. It was alleged that the Arabs learned from the Chinese the art of papermaking from cotton. Various references identified the raw material of paper clearly and unambiguously: papers were called *charta bombycina*, or *bombycis*, *gossypina*, *cuttunea*. What could any of these names mean if not 'cotton paper'? But this is a grave error; all have been deceived, without exception. We must have the courage to admit our mistakes in order to find the new path to the truth.

As I mentioned, Julius Wiesner showed us the way when he demonstrated conclusively on the basis of microscopic examination, as well as by using historical criteria, that paper made of raw cotton never existed and that both Oriental and European papermaking starts with rag paper.

3

The historical development of papermaking from rags (using Arab sources)

Very little was written about papermaking by the Arabs.(ii) However, if all the intriguing pieces are put together, they give a picture of a flourishing trade in paper, the very medium which was to boost the creativity of the Arabs. Nevertheless, the Arab sources, scanty though they may be, are much more plentiful and informative than the Western sources.

Some improbable tales are told about the early days of Arab papermaking.[37] Wattenbach writes 'the production of cotton paper is supposed to have been practiced in China for a long time, and to have been known to the Arabs since the conquest of Samarqand in 704'! This introduction to the subject of paper history is wrong in all respects. There is no evidence that the Chinese used to make paper from raw cotton – which, by the way, is not a suitable material for the purpose. Cotton is not one of the many raw materials for papermaking mentioned in their records.[38] This is hardly surprising as the cotton plant was completely unknown in China at this early time; it is supposed to have been introduced from Ma'bar in southern India during the reign of Qūbīlāy Khan (1257–1294).[39]

It is clear that the story of Chinese cotton paper is completely without basis; but as a theory, it is old and widespread.[40] Wattenbach is only its best known exponent. It is probable that medieval Europe misinterpreted the name *chartis bombycinis*, and a great deal of confusion was caused by this term as a result, as will be shown.

All reports agree that the Chinese were masters of papermaking, but they differ on the date of its introduction to the Arabs. The earliest date for the introduction of paper is 30H (650/1AD) in a reference that claims that paper first arrived as an article of trade from China to Samarqand, the capital of Soghdiana.[41] The preciseness of the date cannot be guaranteed, but it does fall into the period of Chinese influence in Transoxiana.[42] So far, the only evidence on the further distribution of paper is to be found in a confused account by the Orientalist M. Casiri.[43] His authority is 'Ali ibn Muhammad al-Fārisī who claimed that the Arabs first obtained paper when they conquered Samarqand in 85H (704AD). A certain Yūsūf is supposed to have taken paper to Makka where it was alleged to have been made for the first time in 88H (707AD).[44] This particular date is based on a note from the Makkan 'Abd 'Alī Muhammad al-Ghazālī: *Awwalu man sanafa al-qartāsa 'Umar fī Makka sanata thamānīa wa thamānīna* which means according to Casiri's translation: In 88H (706AD) a certain Josephus, family name Amru, was the first to invent paper in the city of Makka and so introduce it to the Arabs.[45]

One look at the Arabic text proves the unreliability of the translation. In the text, the so-called 'inventor' is called neither Yūsuf nor 'Amrū, but 'Umar, who could not have had Yūsuf or 'Amrū as his second name. The confusion becomes greater when one reads the following description from the Persian dictionary *Burhān-i Qāti'*[46], published in 1818 in Calcutta: 'The paper which was made from scraps of raw silk was used in China for a long time before it appeared in Samarqand in the year 30H, and so started its spread from there. 'Ali ibn Muhammad al-Fārisī, the author of a book on Arab history writes: "In the year 85H, Samarqand was conquered by the Muslims, and Yūsuf ibn 'Amrū learned the art of papermaking. On travelling to Makka he taught the art to others. At the time what was used to make the first paper was only cotton"'.

This quotation was also printed in Vuller's *Persian-Latin Dictionary* II, 720, fortunately only in the original; a translation would surely have caused the greatest damage. Supporters of the cotton theory, including their latest representative, Cesare Paoli, would have used it as conclusive evidence.[47]

This whole delightful story with all its decoration comes from Casiri (d. 1791), as the editor of *Burhān-i Qāti'* admits! In it he mixed true and false, original and borrowed ideas. What was reported by al-Ghazālī appears as if it stemmed from the pen of al-Fārisī. The only correct form of the name is Yūsuf ibn 'Amrū. Casiri's reasoning now emerges: he tried to associate 'Umar with Yūsuf, so changed 'Umar to 'Amrū and made this into a second name for Yūsuf because 'Yūsuf son of 'Amr(u)' as mentioned by al-Fārisī did not fit! But the effect of combining the two names led to the assumption of a three-year apprenticeship for Makka's paper manufacture. This is how Casiri's mind worked.

The conclusion is that the reports of al-Fārisī and al-Ghazālī are not connected; it is evident enough that 'Umar could not have invented in 88H the same paper that Yūsuf brought to Makka in 85H.[48]

The transformation of cotton into paper is an invention by Casiri. Further, al-Ghazālī's words cannot possibly refer to the making of paper. The author writing after the event (13th century) passes on a faulty text. The use of the word 'qartās' instead of the proper word *al-kāghid* proves this, and the verb *sanafa*, which does not fit its translation 'invent', is probably a copyist's mistake for *sarafa*. In my opinion, the original quotation was: 'the first person to use paper for writing was

'Umar in Makka' etc. Without doubt this particular 'Umar was not some otherwise insignificant personality, but was the Second Caliph, who was in Makka in the year 88H. The simple mentioning of the name 'Umar in connection with this date is proof enough, for otherwise the chronicler would not have failed to add a patronymic to 'Umar, as was the custom. This was a significant 'first'; by his use of paper while in Makka, 'Umar had shown a preference for this writing material over the others available. This, in effect, sanctioned its use and from then onwards its use spread generally across the Arabian peninsula. Without doubt, this paper was of Chinese origin. If Yūsuf ibn '*Amr*' really did bring samples of paper to Makka, it must have been with the intention of finding a market for it. Paper had been an item of trade for some time. The author of the *Fihrist* reports that he had seen, among other things, a copy of a book by the grammarian, Abū al-Aswad (d. 69H, 688AD).[49] The author understood that this manuscript was written on four sheets of Chinese paper. There were other Arabic manuscripts in that collection written on the same material. Paper was not made in Arabia in the 8th century. When it did start, Makka was certainly not the place it was first manufactured, let alone invented. But I will come to papermaking in Arabia later.

The second great truth does not look any better on close examination: that the conquest of Samarqand in 85H (704AD) is when paper was introduced to the Arab world. It is surprising how long this story managed to survive. The Arab chroniclers, who never fail to give a fine crop of dates as corroborative evidence, do not mention even a temporary occupation of Samarqand in the year stated by 'Alī ibn Muhammad as that for the introduction of paper. What is known is that in the year 85H the Tarkhun of Soghdiana was in full control of Samarqand and living quietly and peacefully.[50]

The first conquest of this ancient and famous town took place in 56H (676AD) under the leadership of Sa'īd, son of Caliph 'Uthmān, appointed Governor of Khurāsān by his father. It was not very spectacular as a conquest, however. There was an amicable agreement which allowed the Arab Commander-in-Chief (and whoever else wished) to enter the city through one gate and leave through another.[51] The Arabs had to battle hard for many years before the Turks in Transoxiana finally submitted to Islam. The rich city of Samarqand with its arts and crafts remained closed to the looting armies even though, in theory, it had been conquered on various occasions. Its inhabitants proved to be tough opponents and brave fighters who managed to shake off the foreign yoke time and again.[52] But the fate of the town was sealed when the brave Qutaibah ibn Muslim set out with the intention of completing the conversion to Islam of the people of the Oxus. The conquest and occupation of the Soghdian capital did not take place until 93H (712AD). It stayed permanently in Muslim hands from then on because the victorious Commander-in-Chief left his brother 'Abdullah with sufficient men in the town to enforce his instructions on the suppression of any opposition.[53]

Considering the circumstances prevailing at the time, there can be no question of trade between these two hostile nations. The Arabs could not have had paper from Samarqand until 712AD, but even so, there is no documentary evidence of it happening then. We can, therefore, dismiss Hammer-Purgstall's statement when he links a paragraph in al-Tha'ālibī's *Book of Morals* on the geographical spread of paper with the year of the initial conquest of Samarqand, 676AD.

Hammer-Purgstall writes: 'The paper of Samarqand is famous because paper first arrived in Persia and Arabia coming from China by way of Samarqand. It happened when Ziyād ibn Sālih led the population of Samarqand away as prisoners after the Battle of Atlakh. Samarqand was not conquered until 56H. This means that the date for the introduction of paper, ascribed by Casiri after an Arab author to the year 30H, is moved 26 years later.'[54]

No Commander-in-Chief with the name of Ziyād ibn Sālih is known from the histories of the Era of Conquest. Neither is a Battle of Atlakh ever mentioned.

Al-Tha'ālibī relates the same story in another work, but in a different version, which shows that Hammer-Purgstall was not very particular about his sources. The paragraph in question reads:

> Among the specialities of Samarqand is paper. This has pushed aside the papyrus rolls of Egypt and parchment, the writing materials used previously, because it is more beautiful, convenient and fit for its purpose. It comes only from there and from China. The author of *The Book of Routes and Kingdoms* reports that paper reached Samarqand from China through prisoners of war. It was Ziyād ibn Sālih who captured them, and there were among them some who could make paper. From then on, papermaking increased and was soon an important article of trade for the inhabitants of Samarqand. In this way, it acquired great value for mankind in all countries of the world.[55,56]

Al-Qazwīnī writes in a similar vein:[57]

> There are many elegant things in Samarqand which travel to other countries from there. The author of *The Book of Routes and Kingdoms* says that prisoners of war brought the secret of paper from China to Samarqand. Included among them were some men who knew the art of making paper, and they were put to work there. From this time on, paper increased in importance until it became a commodity of trade which the inhabitants of Samarqand exported to all countries.

These are valuable references and of the greatest significance; they make it possible to fix precisely the date of the first appearance of papermaking in Islam.

While the eastern areas of the Caliphate were rebelling against the first 'Abbāsi, political changes were taking place in neighbouring Turkish lands which the Arabs were not able to ignore – this applied particularly to Chinese influence in Turkish lands that were fighting each other. The resulting challenge to aspiring Muslim supremacy forced the Arabs to intervene. This is what happened on the occasion we are considering. A war broke out between two neighbouring Turkish princes, the Ikhshi of Farghāna and the ruler of al-Shāsh, known today as Tashkent. The Ikhshi were not strong enough to resist, so they asked the Emperor of China for assistance which he granted. Strengthened by a large number of Chinese soldiers they managed to beat the ruler of al-Shāsh who then had to accept the supremacy of the Chinese as well. This is obviously the reason why the 'Abbāsi emissary to Khurāsān, 'Abū Muslim, sent the Deputy Governor of Samarqand, Ziyād ibn Sālih, at the head of an army against the two Turkish rulers. The bloody conflict took place by the Taraz river.[iii] The fight ended with the total defeat of the Infidels. After heavy losses, they fled across the Chinese border leaving many prisoners behind. This happened in the month of Dhū al-Hijjah 133H (July 751AD).[58]

The story, retold by al-Tha'ālibī after an account by his very old informant, must be historical fact. The name of the locality of the battle coincides: Atlakh is a very important town nearly as big as Isbījāb, the capital of the province. Atlakh was near the town of Taraz which lies to the north of Tashkent on a river of the same name.[59] It was famous for its musk.

This sheds light on some words in the *Fihrist* which might otherwise have cast a shadow of doubt over the accuracy of its excellent and conscientious author.[60] They read, 'Some say that paper came into use under the Umayyads; others claim that it was at the time of the Abbasids!' This statement shows that the change took place just about the time of change from Umayyad to Abbasi dynasties!

As a result we can assume as an historical certainty that 751AD was the year and Samarqand the place of the first papermaking in Islam.

4

The first papermakers

What was the nationality of the first papermakers? The reports leave no doubt that they were not captured 'inhabitants of Samarqand', Soghdian Turks led away by Ziyād ibn Salīh, but were prisoners of war brought back to Samarqand from elsewhere. We can draw the conclusion from Arabic texts quoted previously, that there must have been Chinese papermakers among the prisoners of war. The *Fihrist* makes it even clearer 'they were Chinese workmen who made it (the first paper) in Khurāsān'.[61] Parallels to this can be found wherever wars have led to great population shifts. When, in the middle of the 4th century, Shapur II, son of Hormizdas, marched through provinces lost to the Romans, he is said to have resettled a number of captured inhabitants of Mesopotamia in Susa and other cities of Persia, just as Darius did to the Milesians of Asia Minor.[62] Because there were so many skilled weavers among them, the Arabs credited to that time the velvet makers of Susa and the satin weavers of Tustar.[63]

For the same reasons, the Norman King Roger of Sicily abducted Greek silk workers in 1147.[64] Or to recall a more recent example, we know that Arab prisoners of war were held as slaves in Constantinople in the 9th and 10th centuries and made to follow their previous profession. 'But', adds the chronicler, 'clever is he who does not admit to his craft when he is asked'.[65]

So it is certain that at first Samarqand paper was made by the Chinese method. The material used consisted of various grasses and plants.[66] In fact, the Chinese started making a new type of paper as early as 649–683 during the reign of Emperor Kao Tsung. It was made from a kind of hemp fibre (so-called China grass, the bast fibres, of *Boemeria nivea* (Urtica)). It produced an especially durable paper used for official documents, and after being perfected in 715, was used for the copying of the Emperor's decrees.[67] The usual raw materials used in China, then as now, are the bast fibres of the paper mulberry and young bamboo shoots.

This paper from Samarqand, originally known as Chinese paper, soon found a wide market which raised the commercial importance of the town – not without reason was it said of the people of Khurāsān that they were so skilled in the arts that their country might be a piece of China itself.[68] They were good apprentices of their Chinese teachers and for quite a long time managed to hold a monopoly of the craft of papermaking. No paper was made outside China with the exception of Samarqand but as soon as the Arabs learned the technique, the knowledge spread quickly across the whole world.[69]

5

Samarqand (Khurāsān) paper

This new product, which became famous throughout the Islamic world under the name 'Samarqand paper' or by the geographically wider term 'Khurāsān paper', represents a great step forward in the making of paper pulp.[70] It was a victory of foreign ingenuity over the inventiveness of the Chinese. As soon as they had learned the basic principle of felting paper – that is, making a paper pulp of finely separated fibres and ladling it onto the paper mould – they started to use rags to make the new writing material. 'Khurāsān paper', writes Muhammad ibn Ishāq in the *Fihrist* in 987, 'is made of linen'.[71] The word *kattān* has been translated as flax or linen in this intriguing quotation. There is no doubt that both are correct as al-kattān is *Linum usitatissimum*.[72]

First, from the technical point of view, it is hard to imagine the use of the raw plant material rather than linen fibre made supple through use. Secondly, from the economic point of view, there is no evidence of the cultivation of flax in Khurāsān, including Transoxiana, at the time; if it had been, production would not have been sufficient for this commercial purpose. Of relevance is the proverb of al-Jāhiz (d. 869 AH): 'Everyone knows already that Khurāsān is the land of cotton, but Egypt is the land of flax.'[73]

This is confirmed by many Arabic books on geography in their chapters on the products of individual countries. From the point of view of language, the translation of *kattān* as linen poses no problems.[74]

In this case, *kattān* is not the raw fibre but flax in a processed form (this is analogous to paper 'from hemp' which will be described later). Again, the term does not mean that raw hemp fibre was used, but material manufactured from hemp. The term *min al-kattān* corresponds to the *de pano* that appears in the *Leges Alfonsi* of 1263.

Two conclusions can be drawn from the previous remarks. The first is that it is striking and significant that in cotton-growing countries, the only material described as being used for paper is linen. This means that the Khurāsānians must have found not only raw cotton but also cotton rags unsuitable for papermaking. They must have realised that the natural elasticity of the cotton fibre made soft, spongy and water-absorbent paper. If not properly sized, it was just like blotting paper and unsuitable for writing on. The oldest Islamic papers in the Rainer Collection were hardly sized at all. Therefore, the theory that linen paper developed from cotton rag paper is untenable.

The second conclusion is that the papermakers of Samarqand did not have the flax plant and so went at once to the one source of flax fibre that presented itself – old worn-out cloth, making as it does such fine, smooth and dense paper.

6

Rag paper

Which nationality deserves the credit for inventing rag paper? The question is easily answered in the negative – neither the Italians nor the Germans. The latter, however, seem likely to have discovered rag paper for themselves in the 13th century.[75] Wiesner has now shown this by microscopic analysis of the older Arab documents in the Rainer Collection, and it has also been confirmed by historical research.

Obviously, rag paper was made in the Orient in much earlier times. However, this great advance of such consequence cannot be ascribed to China, even though it is the home of paper.[(iv)] There is some evidence that papermaking from rags started in China in 940AD; by that time, rag paper was already used all over the Arab world, and its production was a flourishing art.[76]

The two candidates can, therefore, be only the Arabs and the Persians.

The fact that the Persian word for paper, *kāghad* or *kāghid*, originated in Samarqand and was adopted by the Arabs might lead us to the conclusion that the Persian people were deeply involved in this developing industry. We question the etymology of the word given by the Persians themselves: *kāghid*, they say, developed from *kāgh*, noise or sound, and the suffix 'd' which describes the rustling sound of paper. This is definitely a contrived explanation.[77]

It is much more likely that *kāghid* is a Persian loan word which can be traced back to a Chinese word for paper. The first Chinese papermakers in Samarqand moved among a Persian-speaking population; the military influence of the Arabs lasted for only the early decades of the occupation.

The period between the first appearance of paper in Samarqand in 751 and its manufacture and dissemination by the Arabs can be considered the first, or Persian, period of papermaking in Islam. There was no Arab involvement until the end of the century, as I will show shortly. In all probability, the Persians started the making of paper in Samarqand and initiated the use of linen rags. Samarqand paper was supreme in Eastern markets from its first manufacture until the end of the Middle Ages, and its reputation was carried far into the Western world. This was certainly the case in the 10th century, as mentioned in the contemporary books on geography by al-Istakhrī, Ibn Hawqal and al-Muqaddasī. The latter praises the incomparable paper of Samarqand.[78,79,80,81]

We can list the most widely used types of papers:[82]

1. *Al-Fir'aunī*, Pharoah paper; probably so called because it competed with Egyptian papyrus, claiming to be similar in construction and appearance (double-layered, similar-sized sheets being joined to make rolls, see later).
2. *Al-Sulaymānī*, probably named after Sulaimān ibn Rashīd, the Khurāsān Director of Finance during the time of Caliph Hārūn al-Rashīd, 170–193H (786–809AD).
3. *Al-Ja'farī*, after the famous and powerful Barmaki vizier, Ja'far ibn Yahyā, died 187H (803AD), who introduced the use of paper to government departments.
4. *Al-Talhī*, named after Talha, the son of Tāhir of the famous Tāhiri family. He was the second governor of Khurāsān from 207H to 213H (822–828AD).
5. *Al-Tāhirī*, after Tāhir (II) ibn Abdullah, the ruler of Khurāsān 230–248H (844–862AD).
6. *Al-Nūhī*, after Nūh I ibn Nasr, the Sāmāni ruler of Khurāsān and Transoxiana 331–343H (942–954AD).

During the 11th century, papermaking reduced in quality as well as quantity so that papers from other areas, such as Syrian paper made in Tripoli, were now considered superior to paper from Samarqand.[83] Egyptian papers excelled in their fineness and smoothness.[84]

However, even in the following centuries, Samarqand paper was praised above that of all other papermaking regions, and *kāghid Samarqandī* continued to be known all over Persia.[85,86] Papermaking was subject to change here as in other areas, and excellent varieties continued to be produced with names like 'Samarqand Sultan paper' and 'Samarqand Silk paper'. The latter did not contain anything of animal origin; it was made from linen rags, and was so called because it was very thin, soft and silky. It continued to be made after the Middle Ages. There were also pure white and coloured kinds which enjoyed an excellent reputation throughout Persia.[87]

The weak cohesion of these silk papers was improved with the help of soap, and smoothness by using glass polishing stones as in olden times. Concerning this latter type of paper, we may assume that in the late Middle Ages, Samarqand was again influenced by Chinese methods. The city always maintained strong trade links with China where the name Samarqand was well known. In Yun-shi and other Chinese works of the Mongol period, it is called *Sie-mi-sze-kan*; it was also called *Sun-sze-kan*, and on a Chinese map of 1330, it is *Sa-ma-rh-kan*.[88] Other names are

Ho-chung-fu (the town between the rivers) and *Ho-fu*. The excellent Ch'ang-ch'un who travelled through western Asia from 1221 to 1224 tells of Chinese workmen living there.[89] Indeed, Chinese influence expressed itself in various branches of the arts and crafts in Samarqand and along the whole eastern border of the Caliphate.

7

The spread of papermaking

Paper had spread beyond Samarqand by the time it took on real economic importance. The rapid spread of papermaking among the Arabs, although encouraged at first by the elaboration of the administrative network, was closely connected to the blossoming of intellectual activity.[90] The famous historian, Ibn Khaldūn (d. 1408)[91], wrote: 'In the early days of Islam, parchment made from animal skins was used for writing scholarly works, letters from the ruler, letters of credit and other official documents. This was possible because the people lived in luxury.'[92] Intellectual activity was small, as we will show later on, and official documents were only rarely issued. Parchment was used to give these documents greater prestige as well as to ensure their durability and authenticity. But soon the output of scholars rose to a flood, and written work produced by the ruler and his government increased so much that the supply of parchment was no longer adequate. Al-Fadl al-Yahyā suggested that paper should be made, and so it was. This new writing material was used by the ruler for his letters and following this, paper came into general use in government offices and for scholarly works. Paper now reached a new level of quality.

The introduction of papermaking to Baghdād came at a later date. As a consequence, the new residence of the Caliph became truly a 'City of Peace'. The date is confirmed by al-Maqrīzī quoting a well-confirmed source that great numbers of papyrus rolls were used in Arab government until the fall of the Umayyads.[93] The first 'Abbāsi, Abū al-'Abbās al-Saffāh, appointed the Barmaki, Khālid ibn Barmak, as his vizier; under him the use of papyrus decreased and parchment was used by preference. It was not until Hārūn al-Rashīd appointed the previously mentioned Ja'far ibn Yahyā to the administration of the Caliphate that a more permanent change took place with the introduction of paper to the government and the people. Apart from the high cost of parchment, the real reason for the change was, as another historian reports, that writing on parchment or papyrus could easily be scratched out or washed off and then altered to deceive.[94] This was not easily done to paper without leaving tell-tale signs.

So the two Barmaki brothers, both of them highly esteemed by the Caliph, were involved in steps that progressively moved the Arab administration towards the use of paper. Al-Fadl is supposed to have advised the change, and Ja'far to have put it into practice. The involvement of these two names with this important event must have an historical basis because both men held office at the same time. In fact we find that al-Fadl was Governor in Khurāsān from 178H to 179H (794–95AD) while his brother held the position of vizier.[95]

By the beginning of the following year, 180H, Ja'far had already been dismissed from this office and was Governor of Khurāsān for only 20 days before being appointed Commander-in-Chief of the Guard.[96] Al-Fadl might have come to know from personal experience the practical value of this cheap writing material that had been used for more than 40 years by the administration of Khurāsān. Similarly, he might have suggested its introduction to the central administration of the Caliphate where his eminent brother Ja'far was in charge. Ja'far's great achievement was to get all sections of the community to accept Khurāsān paper. We have seen earlier that one type of Khurāsān paper carried his name as an honorific for his part in popularising its use.

In short, it is fairly certain that paper was made for the first time in Baghdād about 794–95. From this second site in the Caliphate, the new writing material was to spread much further.

8

The main centres of papermaking

Once the Baghdād government started to use paper, we are told by the Oriental authors that paper manufacturers sprang out of the ground, resulting in a boom in papermaking. Unfortunately, we are often given only hints of the place names. Despite the scarcity of information, they build up a picture of a vast blossoming industry. As with other branches of Oriental arts and crafts such as weaving and ceramics, it would be easier to reverse the question and not ask where was paper made, but rather to ask where paper was not made.

So far we have been able to name only a few localities as main centres of papermaking, using Arabic sources. Here is a list of them and we will discuss some of them in greater detail later:

1. Samarqand.

The mother town of Islamic papermaking has already been discussed. Other papermills were built near Samarqand in later years.

2. Baghdād.

My evidence suggests that paper was still being made here late in the 14th century.[97] The largest size of paper came from here, as we will see later. According to Yāqūt, who spent his youth and his later years until 1213 in Baghdād, the papermakers were located in the Dār al-Qazz quarter around the turn of the 12th century.[98] The Dār al-Qazz, or Hall of Silk, was so called because it was where silk was manufactured; it was woven into striped 'Attabī cloth (tabby [ed.]) in the neighbouring district of the 'Attabīyīn.[99]

3. Tihāma, the south west coast of Arabia.

This is probably the third oldest papermaking region of Islam. It was not long before Tihāma paper was competing successfully with paper from Khurāsān; this is mentioned specifically.[100]

4. Yemen, the land around the Tihāma and the stronghold of the age-old Sabean culture with its famous capital, San'ā.[101]

The high standard of Yemeni papermaking and bookbinding in the 10th century will be discussed later.[102]

5. Egypt.

The conditions for papermaking were more favourable here than in any of the other provinces. Being a flax country, it produced enormous quantities of *linum usitatissimum*. The seed was sown in Pharmuthi (27th March to 25th April) and had reached ripeness by Hathor (26th April to 25th May). Six *sheffel* of seed produced about 30 bales of straw flax to the feddan (6034.18 sq. m).[103] The rent for a feddan of land in ancient Middle Egypt was between 3 and 5 dinars depending on the quality of the land.[104] Mechanical methods such as pressing, beating, rubbing, etc. were not the only means of converting bast material into spinnable fibre – biochemical degradation of the stems by rotting preceded mechanical treatment. There were sites in all areas of Egypt for this 'retting', or fermentation, but they were especially numerous in the Fayyum. The best flax was that of Giza. Strict police supervision ensured the Egyptian flax was not mixed with any from Nablus or Syria, an offence judged as fraud.[105]

It is evident from many documents in the Rainer Collection that all classes of the population derived a high standard of living from this flourishing cultivation of flax.[106] It is not surprising that Coptic weavers made Egypt supreme in the craft of linen weaving, and their high-quality product was traded internationally. The common use of linen in Egypt provided rags which, together with the enormous amount of linen taken from graves and rubbish dumps, were the ideal material for papermaking.[107]

If paper was being made in the 9th century, it can only have been on a small scale. Although paper was used then, it was not necessarily an Egyptian product. The oldest papers in the Rainer Collection could just as well have come from Khurāsān, Baghdād or Arabia. An example is the roll of paper from which a craftsman cut an architectural form of the projected minaret of the great Ibn Tūlūn Mosque, then under construction in Fustāt (Old Cairo) 876–878.[108]

There was a change by the 10th century with a great increase in the quantity of paper in Egypt. Al-Tha'ālibī (961–1038) praises the quality and smoothness of his native product, and his praise is confirmed by a particularly fine, thin piece of 11th century paper in the Rainer Collection.[109] It bears the inscription 'an Egyptian paper', and though it has an area of 75 sq. cm and was probably coated with earth, it still weighs only 0.375g.

When the traveller Nāsir-i Khusraw visited Egypt from 1035 to 1042, he saw how, in the bazaar of Old Cairo: 'They give a container of glass, pottery or paper with everything they sell in the greengrocers, perfumers and haberdashers in the markets there. In brief, there is no need for the purchaser to take a container.'[110] This must have been a strong wrapping paper, certainly locally made, that was still being made in Egypt around 1200. The famous Baghdād doctor 'Abd al-Latīf, travelling through Egypt, recounts 'how the bedouin and fellāhīn used to search the burial chambers for mummy wrappings. If they were strong enough, they were used for clothing; if not, they were sold to papermakers who made wrapping paper for the spice traders.'[111]

Unfortunately, we have no detailed information about where Egyptian paper was made, even though 'Egyptian paper' was an important item in the administration of the Baghdād Caliphate.[(v)] We have valuable details of the size, the names and the uses of this paper up to the late Middle Ages. We must leave unanswered whether the old papyrus-making areas of the coastal town of Bura in the Damietta district of the Delta, or of the Fayyūm, were converted gradually to the making of paper.[112] We can assume, however, that Cairo was the Egyptian centre of papermaking and of the paper trade. The Old Papermakers Street crossed the Turks Way which led up to the al-Azhar Mosque.[113] The Papermakers' *Khān* lay between Bahā' al-Dīn Street and the small market place of the Amīr Juyūsh district to the east of which was the Cauldronmakers' market. This *Khān* consisted of a collection of lodgings and a papermill, and was on the site of the Stables of the Pages.

6. Syria.

The first paper in this province was made in the capital, that earthly paradise, Damascus, known as the 'Bride of the World', the residence of the Umayyad Caliph.[114] The city was the focus of commercial activity in Middle Syria. Based on reports by al-Muqaddasī in 985–86, we know that 'Damascus paper' was exported to the West as early as the 10th century.[115] *Charta Damascena* was known to the Muslims of the late Middle Ages as 'Syrian paper' and along with other crafts such as weaving of damask, soon came into competition with goods from Baghdād. The papermill in Damascus gradually absorbed the other mills. We will return to this later in connection with the various types and uses of paper from Damascus. I will add that in Ibn Battūta's time (1327), paper, reed pens and ink were sold in the Papermerchants' market near the large eastern gate of the 'Umar Mosque.[116]

Al-Muqaddasī writes that in Palestine by the lake in the old Galilean town of Tiberias there was a second papermill.[117] It had produced and exported paper for as long as the mill in Damascus. It is noteworthy that Tiberias, a prosperous town, had a second speciality besides the preparation of linen. This was the making of ropes and mats from locally grown esparto, *Stipa tenacissima*, an industry which probably had an effect on local papermaking.[118] For a long time, English papermills have used this material mixed with straw to give an especially beautiful, firm and opaque paper.

There was an excellent papermill in the old Phoenician town of Tripoli on the Northern Syrian coast. The previously mentioned Nāsir-i Khusraw writes: 'good paper is made there, similar to the paper of Samarqand but better'.[119] The town was under the rule of the Egyptian Fātimid Caliphs and, Nāsir-i Khusraw assures us, was much visited by Greek, French, Spanish and Maghribi traders. Papers from Tripoli must have reached the West even in his days (1035–1042). A fourth papermill was situated in Northern Syria, a little above Apamea in the fertile Orontes valley: the old royal Hittite town of Hamath was renamed Epiphaneia by Antiochus IV Epiphanes, and the name was changed again by the Arabs to Hama. Hama paper must have been very popular in the early Middle Ages because, after the papermill was transferred to Damascus, it continued to be made under the name of al-Hamawī.[120]

There was a fifth Syrian town which made paper. It is not mentioned in the Arabic texts but I think there are reasons, which I will explain later.[(vi)] It is Hieropolis, or Mambīj in Arabic, situated in an oasis near the Euphrates just off the great military highway which runs from Northern Syria to Mesopotamia.

7. North Africa, al-Maghrib.

We know of only one place in the Muslim part of North Africa where paper was made in medieval times.[121] Fez was founded by the Idrisis in 808, and papermaking must have reached its peak there in the 12th century at the latest. The following historical account proves that paper was already common in the Maghrib for purposes beside writing the Qur'ān and other books.[122] The prayer niche (mihrāb) of the Qarawīyīn Mosque was especially richly decorated, surrounded with coloured and gilded carved arabesques. It was feared that when the puritanical Almohad ruler 'Abd al-Mu'min came to Fez, he would destroy the carvings in the mosque so a cunning deceit was devised to save them. Paper was stuck over the decorations and then covered with plaster to give a smooth surface that could be whitewashed. The elaborate carvings disappeared under this plain surface.[123]

It is obvious that for this procedure a sized paper of medium weight and of considerable strength and resilience was needed. This story, together with the fact that our libraries still have books from the Maghrib made of the finer types of writing paper, justifies the conclusion that a wide variety of papers was made in the Maghribi capital. Although we have no names, there must have been competing mills in Fez. The confirming evidence is that, a few decades later from 1184 to 1213, there were 400 millstones in use for papermaking in Fez.[124]

8. Early Spanish writers such as Ibn 'Abd Rabbihī (860–940) describe paper and its use on the Peninsula[125]; it is known also that the catalogue of the great library of Caliph Hākim II (961–976) was written in booklets, each of 20 sheets of paper.[126] However, the first positive evidence of papermaking comes from a much later date. In 1154, al-Idrisī was the first to praise the town of Xativa as a centre of papermaking. Xativa (Jativa [ed.]), known in Arabic as Shātiba, previously Saetabis,

is known today as San Felipe. Al-Idrisī notes: 'Xativa is a pleasant town with palaces and fortifications proverbial for their beauty. Such fine paper was not made elsewhere in the civilised world, and was exported to the east and the west.'[127] There were considerable exports to the Maghrib and other parts of Africa.[128] The remark by Yāqūt (d. 1228) about Xativa that, 'Excellent paper is made there and exported to other towns in Spain', certainly does not imply that export was more restricted than is indicated in the previous quotation.[129] In those days, a large part of North Africa and Spain was ruled by the Almohads.

I cannot understand why Keferstein says that, 'the Arabs made paper in Africa at Septa (known as Ceuta today) from an early date, and for a long time supplied neighbouring Spain'.[130] I have no knowledge of the source of this information. Also, confusion is caused when Gardthausen makes al-Idrisī report (according to Jaubert's translation) that besides Xativa, Valencia and Toledo were noted for papermaking.[131] They certainly were, but there are no reports of this in Arab times.[132]

Spain was similar to Egypt in that the making of linen paper was stimulated by the availability of considerable amounts of rags from the local linen industry famous since the time of Pliny.[133]

9. Persia.

Papermaking in Samarqand and Khurāsān has already been discussed in detail. We return to Persia prompted by later evidence of an Iranian paper industry. It was mainly in Tabriz, the capital of the district known to the Arabs and Persians as Āzarbaijān. This town, known as 'Little Cairo' by the historian Vassāf (1213), always had a good name based on the skill of its inhabitants. It owed its prosperity especially to textiles, even though they can be shown to be copies of foreign designs. The coloured and striped half silk 'Attābī fabrics, and the splendid Siglatun cloth (heavy woven silk [ed.]) point to Baghdād.[134] The khata'ī materials point to China.[135] When the Mongols invaded the country in 1231,[136] the people of Tabriz surrendered, purchasing the Kh*ā*n's mercy with their precious goods and by making a splendid gold-embroidered panel on satin for him.[137]

'Tabriz paper' was known to the Persians as 'the speciality of Tabriz' and was probably, despite its name, based on a foreign type. In fact, the whole of papermaking in Tabriz suggests a tradition of borrowing. Īl-Khānī paper money was based on Chinese examples and was made for the first time in Tabriz in 1293 whence it was distributed. The Persian paper called *khatā'ī* relates to Chinese paper in the same way as the similarly named silks of Tabriz.

This is understandable because, in Persia and in many towns in Arab and Persian Iraq which were an integral part of, or were vassal states of the kingdom of the Īl-Khān, the output of papermills increased enormously because of the introduction of paper money (according to *Tar'īkh-i Vassāf*, 1293). From among the coloured papers of Persia, *Mukhayar* or 'watered moire', was very popular; it was possibly the equivalent of today's marbled paper. We do not know what *Qāsim-begi* paper was like.

10. India.

We have only names but no descriptions. The general term *Hindī*, Indian paper, covered a colourful assortment. However, we know that Indian 'Silk' paper is clearly related to the silk paper of Samarqand; the types called *Niẓāmī Hindī* and *'Ādilshā hī* sound as if they were fine high quality types of paper. Only the name *Daulatābādī* remains to show that there was a papermill in the town of Daulatābād.

9

Arab papermaking materials

When one looks at the enormous area in which the Muslims made paper, it is essential to ask if linen rags were without exception the only material used. From the historical point of view, the answer must be no; Wiesner came to the same conclusions based both on the evidence of his microscope and of his historical research. The second papermaking material that he found is derived from the bast fibres of *Cannabis sativa*, i.e., hemp rags. The Arab papers in the Rainer Collection came from Egypt, and here much less hemp was used than linen rags. The proportion of linen papers to hemp papers was 3 to 1.[(vii)]

The Arabic sources give more information about hemp paper. According to the *Dīwān al-inshā*, it is made from *nabāt al-qunnab*, the hemp plant, *Cannabis sativa*. 'Paper from hemp' does not mean that new hemp fibre was used for making paper. Far from it. Just as when I write about a paper 'from linen', one has to understand that the hemp fibre used came from discarded manufactured products. The author of the *Dīwān al-inshā* emphasises the point as follows:

> They collect the taller and stronger female plants, taller than the male and the same height as the Persian reed. It is broken and softened in water to give clean hemp fibre. This is twisted into thick ropes such as those used for ships' rigging. When these are worn out, they are sold to papermills for use as raw material for papermaking. The quality of the resulting paper depends on the development of the plant when harvested; only the fully ripe plant yields the coarse bast fibres needed for rope making. It depends on the time of year that it is prepared – it is best in spring (for bleaching to half-stuff) the care with which it is rinsed and the quality of the water

> used for rinsing. It is affected by the degree to which it is macerated through boiling the stuff in lime water and on the humidity of the ground on which it is prepared (the stuff is kept in damp ditches before pounding). Finally, quality depends on the smoothness imparted to the paper by rubbing both sides with a glass polisher.

So much for the *Dīwān al-inshā*.

The custom of using old hemp ropes has continued even in modern papermaking because the characteristic strength of hemp fibre yields a very strong paper – but less fine than that made from linen, of course. I managed to isolate a remnant of hemp cord (or hemp sacking) 17mm long and 1.5mm wide that had escaped the pounding (Rainer Collection Paper 7331). There is no doubt, both from usual papermaking practice and from Wiesner's microscopic examination, that hemp cloth was used. Such cloth has been found in Egyptian tombs.

It would be surprising if the Arabs, who had always had friendly relations with the Chinese, had not adopted some alternatives to rag. With the exception of the method to be described later, the alternative that first springs to mind is that of making paper from the young shoots of bamboo; this plant was known to Arabs and Persians and was imported from India. There is, among the papers from Ashmūnain, a 10th-century list of the possessions of a certain Abu al-Hasan which includes 'a bundle of bamboo'. However, Wiesner's examination of the Rainer papers for rag substitutes has yielded no results. It is simply a fact that rag paper was the easiest and cheapest paper to produce. The Arabs usually kept to the two main categories of linen and hemp paper. Whenever cotton fibres are found, they are there as chance inclusions, perhaps as a result of insufficient sorting of the original rags. There has never been any paper made from raw cotton.

10

The origin of the myth of cotton paper

As far as I can judge after almost 20 years of careful study, the Arab and Persian sources are completely silent about paper that is allegedly made of cotton, *qutn*. The only reference, as I have mentioned, was invented by Casiri. Western sources, however, employed terms which, through common usage, implied that cotton was a raw material of paper. This mistake was widespread; the most recent person to make it is C. Paoli in his essay in answer to Briquet's 'Carta di cotone e carta di lino'.[138] The reasons he advances are false, the conclusions erroneous.

Not wanting to repeat all the references which start with Muratori and Montfaucon and continue to Wattenbach and Paoli, I will deal only with some of the more questionable names for paper. They are: *charta bombycis*, *bombacis*, *de bambace*, *de bombice*, *de bambasio*, *bombycina*, *bambacina*, *bambasina*, *bambagina*, and so on. These terms are not evidence that cotton was an actual ingredient of these papers. On the contrary, there are many reasons why these should have been used as common terms for paper that looks and feels like cotton, particularly when we appreciate that with the development of language, *bambax* and *bombyx* came to be used to denote any fine fibre.[139] I have evidence showing how these terms came into use.

Marco Polo wrote,[140] when dealing with the Chinese paper money: 'He believes that the bark was taken from a certain tree, and from this bark, sheets of paper were made as from cotton'.[141] Similarly, Oderico da Pordenone (about 1317–29) mentions 'a sheet resembling cotton'. The superficial similarity in meaning led Rubruck to the wrong conclusion that 'the common money from Cathay is of paper from cotton'.[142,143]

It was known that Chinese paper money was not made from cotton but from the 'bark' (cortex), or more accurately, the bast fibres of the paper mulberry,

Broussonetia papyrifera. These fibres were soft to the touch, tough and absorbed water easily; in short, they looked and felt like cotton. The similes, used by Marco Polo and Oderico for comparisons were misinterpreted by the inexpert Rubruck as facts. There must be similar examples concerning the existence of linen rag paper in medieval Europe.

We must add to Rubruck's useless information that of Accursius, who explained the two terms *chartae papyri* and *chartae bombacinae*.[144] These were used by Frederick II as having the same meaning because under the name *Charta* are included those things which are made from cotton. When the Consul of the tradesmen of Pisa promised to copy their charter *in carta di bambace*, and the notaries to use paper *di bambace sane*, they were using the expression *di bambace* which had long been the conventional term for paper made from a cotton-like material.[145] The feel of linen rag papers makes this understandable. The surface of Oriental and European papers from this period gives an impression of cotton though this derives from the use of linen rags. From earlier Arab times, good, splinter-free, bleached and carefully combed flax had provided a substitute for cotton for the linings of dresses, paper caps and bookbindings. Moreover, we know that today, bast fibres of flax can be turned into very thin, supple, white, cotton-like material using methods known as 'cottonisation'. It takes only one further step to change the name of the fine, flax paper called *di bambace* to *gossypina* or *cuttunea*; these are both just synonyms of the earlier expression. No informed person would have imagined that actual cotton was used, just as today no one would look for cotton, silk or velvet in our cotton, silk, or velvet papers. For the majority, however, such terms have always continued to be used because of the external characteristic of the papers.

In the course of my detailed study of the question of 'cotton' paper, I have developed a theory based on historical evidence. The term *charta bombycena* is not derived directly from the similarity between the paper and cotton but arrived at indirectly.

My reasons are as follows. I mentioned previously in the list of papermills in Syria one at Hieropolis. This is the old Mabōg, i.e. source, which on transliteration into Latin through Greek, gave Bambyce and in Arabic, Mambīj.[146,147] The Franks called it Bambych. This former capital of the province of Euphrates was built by Constantine the Great.[148] It was a five-day journey from Antioch, three days from Aleppo and one day from the Euphrates.[149,150] In Justinian's times, it was a bishopric, and was conquered by the Sāssāni Persians in 540AD.[151] Under the emperors of Eastern Rome, it became poor and desolate but revived after its conquest by the victorious Muslims as they spread across Syria.[152] Mambīj or Bambyce controlled the military highway running from Northern Syria to Mesopotamia and it became a strategically important fortress for the defence of the Northern Syrian border. Such a strategic objective saw the Christian Cross and the Muslim Crescent changing place regularly on its towers.[153] Nicephoras Phocas brought Bambyce under his control but John Tzimisces captured the town again in 975 after its previous decline.[154] Bambyce was in Christian hands again in the 11th century when

Romanus IV Diogenes wrested it from the hands of Mahmūd, one of the Mirdāsi Amirs, in 1068.[155] The Greeks put an army of occupation into the citadel but soon had to leave the rest of the town due to famine. Seven years later in 1075, Mambīj was back in the hands of the Mirdāsi under Nāsir, son of Mahmūd.[156] But this was not the end of strife; the Seljuq Malik-shāh conquered it in 1086, and it was stormed and looted by the faithless Jocelyn of Courtenay in 1108. When Tancred, the Governor of Antioch, set out to conquer Bambyce in 1111, he found it desolate because the inhabitants had fled in fear of the Franks. The citadel, however, could not be taken by force.[157] The Christians guarded it jealously; they did not want this important town to fall into the hands of the enemy because of the great strategic advantage it bestowed.

For this reason in 1124 they helped Hasān, the ruler of Mambīj against the warlike Artuqi of Aleppo who had set out to conquer the fortress but who was killed during the siege.[158] Their plans were destroyed with the arrival of an even more powerful leader, the scourge of the Christians, Imād al-Dīn Zangī. His triumphant campaign soon (1128) made Bambyce part of his rapidly expanding empire.

The town was recognised as a valued possession, and although it had suffered greatly during these regular exposures to war, its development was not hindered in any way. The great days for Mambīj started when it fell to Saladin the Great.[160] It remained in Ayyubi hands after the division of Saladin's estate following his death in 1193. Even the close proximity of the Mongols in 1260 failed to cause much damage in Bambyce.[161]

All the travellers who visited the town during the period I have described praised its climate, culture and manufacturers. The Persian, Nāsir-i Khusraw passed by the town in his travels of January 1047, seeing no buildings outside the city walls; he probably saw only the agriculture.[162] The Sicilian Ibn Jubair visited it in 1184 and described the huge citadel, the wide streets and the great market places. He comments on the size of the shops, the inns and also on the covered bazaars. Ibn Jubair remarks that the inhabitants are honest and polite; they were straightforward in trade and were generally law abiding.[163] Even al-Qazwīnī (d. 1283) described Mambīj as a prosperous town, with well-appointed cultural facilities, schools and residential areas all surrounded by massive walls. The remains of the excellent system for channelling water were also seen by Pococke.[164,165] Decline came fast: al-Dimashqī (d. 1327) mentions the town without any special comment, but according to Abū al-Fidā (d. 1331) the greater part of the town and its surrounding walls had collapsed.[166,167]

If I am allowed to speculate, I would say that the town of Bambyce, or Mambīj, first gave Bombycine papers their name. Just as the craft of weaving flourished there and its output was called, according to al-Bakrī (d. 1094), Bombycine cloth after the name of the town, so a papermill situated there might have given its output the same name.[168] Their paper crossed the local borders to the west just as did the papers of Damascus. Considering the appearance of the paper, it was not a great step for the *Xartis bambykinos* of the Greeks and *charta bambycina* of the Franks to change from Bambyce or Mambīj paper to *Xartis bombykinos*, that is, 'cotton paper'. This is especially likely when we consider that the town itself derives its name from a similar confusion in Greek, which was translated at one time as 'Cotton town'.[169] At the same time that Bambyce started to decline,[170] Northern Syrian papermaking finished or was transferred to Damascus where Bambyce paper continued to be produced under its old name (cf. the abandonment of the Hama papermill).[171] This must be the only reason why it was '*charta bombycina sive Damascena*' that continued to have a reputation in the West.

If, following this argument, we conclude that in the Middle Ages, the term Bombycine paper was used only to describe the external appearance of this paper, we can find a striking example by way of confirmation. In Chapter XXIII of his famous work, Theophilus wrote concerning the placing of gold leaf between sheets of paper and then: '*Tolle pergamenam graecam quae fit ex lana ligni*'.[172] It is agreed that by '*pergamena greca*' one must understand paper that is parchment-like in colour, smoothness and polish, as described in medieval Oriental sources. Also the second part of the sentence, '*quae fit ex lana ligni*' states that the paper is made of cotton. However, an old scribe would not have believed this because he knew that there was no such thing as 'cotton paper'. There is proof in the Wolfenbuttler Codex, the oldest of the Theophilus mss., where the word '*lini*' is found instead of '*ligni*' which was incomprehensible to the copyist (see for instance Vienna Codex 2527 and the Harleian ms. on which the Hendry edition, London 1847, was based). The Codex Regius in the National Library in Paris, written in 1421, even reads '*ex lana ligni id est papirum*'.

It cannot be assumed that the technically knowledgeable Theophilus, who lived until the beginning of the 12th century, would have thought of cotton when he wrote '*lana ligni*'.[173] One would have to find proof first that '*lana ligni*' meant cotton in the Middle Latin. It seems to me that a different interpretation makes better sense.

The controversial term '*lana ligni*' I interpret not as cotton or as 'flax wool' (*lana lini*), but as another weavable bast fibre derived from the stems of plants. '*Lana ligni*' (wood wool) was a papermaking material that looked and felt like cotton. The bast fibres that form a layer beneath the bark of certain trees, especially the mulberry tree (*Morus alba*) and its varieties are particularly good for papermaking.[174] After washing and bleaching, the fibres were boiled in alkali derived from wood ash to produce a wool- or flax-like fibre that could be easily used for papermaking.

It seems that in the Middle Ages, the bast fibres were usually mixed with the bark (the cortex itself), as Marco Polo described. Possibly this is what is referred to by the misunderstood term *charta corticea*.[175] This expression could possibly be interpreted in certain cases as a paper of true bast fibres of foreign, even Chinese, origin,[176] but *charta corticea* is in no way connected with the Greek word *Zyloxartion* or papyrus.[177] This writing material was made in the same way as rag paper by the felting of tree bast fibres, the medieval term for which was apparently the word we find in *charta zylina*.

It seems fairly indisputable that the Arabs, and after them the Byzantine Greeks, used bast fibres as an alternative to rags for papermaking in Asia Minor, Northern Syria and the south Caspian provinces.[(viii)] They must have learned from the Chinese this second use for the mulberry tree which was grown in these regions for the cultivation of silk.[178] However, in none of the papers that Wiesner studied under the microscope was it possible to identify any.

The connection between the mulberry tree and papermaking was not unknown to the Arabs. No part of western Asia was more famous for silk production than Bambyce, which is why attempts were made to link its name and the Greek word for silkworm (Bombyx).[179] The cultivation of mulberry trees was so extensive that almost nothing else was planted.[180] Perhaps if, one day, the existence of a papermill at Bambyce is proved conclusively, we will find that this readily obtainable alternative was used instead of linen and hemp rags in the factories.

11

The technology of paper

As I explained earlier, paper was originally called by the Persian name *kāghad* or *kāghid* (probably a Chinese loan word) or in Arabic *kāghidh* or *kāghid*.[181,182,183] From this is derived the Arabic *kāghida*, a sheet of paper; *kāghad*, a papermaker, in Persian *kāghidi* or *kāghidgar*; the Persian *kāghidkhāneh*, papermill; and *kāghid furūsh*, paperseller.

Arabic terms for paper all derive from 'waraq', literally a leaf, later to mean paper. For example *qirṭās waraq*, a sheet of paper (plural *awrāq*); from this is derived *al-warrāq*, the paper maker; and *al-wirāqa*, the papermaker's trade.[184,185,186,187]

Many clues suggest that the authorities took over the paper industry from the beginning, confining it to one factory in each area, thus leaving little scope for private enterprise. A manager was put in charge of the paper factory, or *kāghid khāna*, in each area.[188] These factories were connected to where the fermented and macerated rags were further broken down after cleaning and bleaching; this process was achieved by pounding or by milling. It must be assumed that the former method, in which the paper pulp was pounded in stone mortars with a wooden pestle to give a liquid pulp, in the Chinese way, was used at first. However, the latter soon came to be preferred. Attempts have been made to associate the milling of rags with the early days of papermaking. For example, Mordtmann in *Das Buch der Lander* translates the relevant passage in the facsimile edition of the *Liber Climatum* by al-Istakhrī as if the last Sassani king Yazdagird was killed in a papermill in Marw, and treats papermaking as a usual occupation for the year 651 AH.[189,190] Correctly, this paragraph should read 'He, Yazdagird was killed near Marw in a mill in the village of Zarq'. It was a village called *Zarq* not *waraq* (paper) that was situated near Marw on the river al-Razīq, known also as the Zarq river.[191] Nevertheless, in the Orient, papermills are very old – much older than any European mills – and are definitely an Arab invention. This honour Europe must concede to the Arabs although they claimed it for themselves. Water-powered mills were not pioneered in Italy (Fabriano) or in Spain (Xativa); the first one in Germany started in Nurnburg in 1390.[192] Keferstein is completely wrong to say 'We Germans probably did not use stamping machines for breaking up pulp, as did the Italians, who learned the method from the Moors. We used instead our own invention, the hand mill'. Long before the establishment of the first papermill at Nürnberg, we had knowledge of millstones powered by hand or water.[193] The mill was in common use by the Arabs as early as the 12th century. Far from using stamping mills, the Moors had no less than 400 millstones in use for papermaking in Fez during the reign of the Almohad ruler Ya'qūb ibn Yūsuf ibn 'Abd al-Mu'min (1184–98) and his successor, al-Nāsir ibn Ya'qūb (1198–1213). The same chronicler makes clear in another reference that these stones are both hand- and water-powered, and he gives further details.

There is no doubt, either historically or scientifically, that the raw material used in papermaking was almost exclusively rags, except for bast fibres. From my experiments on unimportant fragments of our papers, burning or soaking gave a strong smell of rags. This shows that the cleaning of the rags with lime water was not always carried out thoroughly. The same is true of the mill itself. There were considerable difficulties in the way of thorough milling because of the toughness of the fibres and the inefficient stampers and millstones of the early papermills. It is often possible to isolate whole threads and even remnants of woven material with warp and weft from our Arab papers. I might also mention that several papers show white threads; one has an exceptionally long thread and several have intact pieces of linen.

The next step after papermilling the rags into paper pulp was to size it. It is to Wiesner's credit that he discovered the method of sizing using starch paste. Briquet was wrong when he claimed that they used tragacanth after sizing with animal glue. Wiesner is certain that starch, i.e. a vegetable not an animal size, was the usual method of sizing paper.

Wiesner also discovered that, in addition to this Arab method of sizing with starch, paper was filled using the same starch (unchanged and unmodified starch grains). The size made the paper suitable for writing upon, while filling made it white.[194] Wiesner identified the starch grains in our papers showing that sizing was achieved with wheat starchpaste and filling with wheat starch granules.

Fortunately, I can now add historical evidence to these results.[195] There is proof that from the 10th century the Arabs knew how to make wheat starch as used in this process. According to al-Muqaddasī in Yemen, starch was used exclusively for sizing and filling: 'and in the Yemen they stick their sheets of paper together and line their book covers with wheat starch'.[196] I will talk later about this joining together of sheets of paper and will only remark here that the boards used for bookbinding were made from ladled pulp and were not pasteboards made from layers of paper. The pulp for 'ladled board' had to be thicker than for making paper, so a large amount of wheat starch was added to increase the strength and weight of the board, as in present-day practice. Thus it becomes clear that wheat starch was also added to make heavier and more substantial papers.[197] It is interesting that in some other regions, such as Palestine, *ashnās* paste made from the root of the asphodel was used, this method being unknown in the Yemen.[198] In Egypt both substances were known as trade commodities. Wheat and asphodel starch were much used in various crafts and subject to strict police regulation.[199] Asphodel starch was often used in the prohibited practice of making silk thread and coloured linens heavier, or to dress silk with a gloss. The restriction on wheat starch was lifted in Egypt by an edict dated Thursday, 6th May 1387.[200]

The next stage in papermaking is the moulding of the pulp to give sheets of paper. The early Arab papers in the Rainer Collection show a high standard of craftsmanship in papermaking; this applies not only to the 9th century but possibly even to those of the 8th century. The high standard comes from the wire mould which, it was assumed until recently, was used only in Western papermaking starting in the 12th century.[201] Our papers prove that soon after they began to make paper, the Arabs ladled pulp from the vat onto a mould[(ix)] covered with wire mesh and so were able to make true laid paper! Careful study of these papers together with a lot of measuring shows, as far as I can see, that three different types of mould were used by the Arabs:

a. Moulds with parallel, straight lines lying across chain lines.
b. Moulds as above but with no visible chain lines.
c. Moulds with very fine wire mesh similar to our wove moulds.

The first two moulds were used to make laid paper of low or medium quality and of uneven thickness; this paper when held to the light clearly shows translucent lines from the impression of the mould.[(x)] The third type of mould was used for much finer paper, requiring effort and time for its making. It gave sheets of an even thickness which showed no lines after pressing. Close inspection revealed only traces of the wire mesh when the thinnest pieces were examined by transmitted light. I will explain later the use made of the thinnest of the Arab wove papers.

There was a gap of 1mm between one wire and the next so that there were 6 wires/cm; for example, Paper 6059 is 15cm long and has 90 lines.[(xi)] Only rarely (Paper 7348) could I count 5 wires/cm.

The sheets were sized and dried after removal from the mould; they were then smoothed individually with glass polishing stones. The reason for the difference between the sides of Arab papers – one side being smoother than the other – might have been the method of couching or stacking the freshly made sheets. The wet and extremely soft sheets might have been laid in the Chinese way onto a wall, the surface of which was heated by fire or the sun.

Alternatively, they might have been laid onto coarse felt to which surface it adhered more strongly than to the wire of the mould. To produce papers with both surfaces smooth and suitable for writing, two sheets were stuck together.[(xii)] We are not without historical evidence for this practice; the significant paragraph by al-Muqaddasī has already been quoted: 'In the Yemen, paper sheets were stuck together with wheat starch'. One can indeed see from the papers in the Rainer Collection that most of the papers have been made in this way. Although the pasting was carried out carefully and is not obvious from the surface, the two sheets are by no means inseparable. I have been able to separate some quite instructive examples without doing any damage to the text. After separation, the inner surface always looks rough, woolly and felt-like.

The two-layered papers were usually opaque like the single sheets made from thick pulp. For these two-layered papers, sheets of different thickness were used; the thicker was always the laid one, and the thinner was always the wove. Only the better papers retained their transparency sufficiently that when held to the light, one could just see the laid lines. It is worth noting that one could write on both sides of even the thinnest paper if it was produced in this way.

The Persians called these double smooth-surfaced papers 'two layered'.[202] From the very beginning of the Middle Ages up to the 15th century, it is not possible to find an Oriental paper, which, if it is strong, is not double layered.

At first the resulting sheets were small due to the method of ladling pulp, and were cut to different sizes. In the 10th century it was said of the 'Sulaiman paper' made in Khurāsān that every sheet had space for 20 lines to a side.[203] If we understand these to be codex sheets, we can deduce that a book size was about 24 x 16cm, and therefore, a whole sheet was about 32 × 24cm. Usually a single sheet, or one made by sticking several sheets together, was known as *darj* (plural *durūj*); a page was *qartās* = *charta* (plural *qarātīs*). In the Commentary of al-Harīrī we have 'the *qartās* is cut from *kāghid*'. We also have *kāghid*, *waraq* and *sahīfa*.[204,205]

It was not possible to make larger sheets until centuries later. The best large format sheets came from Baghdād, its rival papermill in Damascus and from the Egyptian papermaking areas. In Egypt they followed the Baghdād fashion. The first large size 'Complete *Tūmār*' had a width of 73.329cm and was made during the reign of the Mamlūk Sultan Shaikh al-Malik al-Mu'ayyad in the year 815H (1412AD). I have learned from experience of Oriental manuscripts in European libraries that the width of book pages on average is two-thirds of their length. These largest sheets must have been about 109.9cm long because this would have folded to give a book page 73.3 × 36.6cm. A splendid Qur'ān, now in the Municipal Library in Leipzig, was taken by Saxon troops during the conquest of Budapest in 1683.[206] It was written in Baghdād in 1306 for the Great Khan Uljaitu and has a trimmed page size of 66.2 × 47.7cm. This means that a full sheet of the largest Baghdād paper (undoubtedly the type used here) must have been 95.4 × 66.2cm. The 'Complete *Tūmār*' made in Egypt was indeed similar to that produced in Baghdād. Another example, the Codex CCXXXVII from Damascus now in the Royal Library, Copenhagen, was written in 738H (1337AD) and measures 22.1 × 15.5cm. Here again, the width is two-thirds the height, so the size of the full sheet was 88.4 × 62cm; therefore, these were folded from sheets of the so-called Incomplete

Baghdād size providing one-sixteenth pages for the codex, that is, eight sheets folded once (*mudarraja*), or bifolios (*kurrās*).(xiii)

Of course it must be borne in mind that measurements taken from manuscripts are smaller than the original size due to trimming. By using 1½ ells (73.329cm) given for Egyptian 'Complete *Ṭūmār*', we have a safe starting point for the size of the uncut sheets straight from the mill.

The results are:

a. Baghdād whole sheet	109.9 × 73.329cm
b. Baghdād half sheet	73.329 × 48.8cm
c. Complete *Ṭūmār*	109.9 × 73.329cm

Paper of the largest size was called *qaṭ'*, piece or section, or *qaṭ' al-waraq*, 'a piece of paper', the same as *tomarion*.[208] Where there is need to describe size further, it is called 'the large *qaṭ'* ' or 'Complete *qaṭ'* ' for the whole or largest sheet and 'half q*aṭ'a*' for a half sheet of paper. In the autobiography of the Mamlūk Amīr Shaikhu (d. 1351), a noted penman, one reads that he gave wonderful pieces of calligraphy 'on large Baghdād *qaṭ'*' to the Umayyad Mosque.[209] In the year 1330, a sealed letter from the īlkhāni Sultān Ghāzān Mahmūd arrived in Damascus 'and it was written in a crude Mongol script on a piece of Baghdād paper'.[210] The famous Īlkhāni vizier and historian Rashīd al-Dīn (1240–1318) completed his great Persian work of history *Jāmi' al-Tawārīkh* in 1310(xiv) and reports how copies were made 'on paper of ultimate quality and beauty in the large Baghdād format and in a large legible hand'.[211,212] In other references one reads 'the book was composed of 60 sections of Complete paper'[213]; also 'as much Complete paper as was stored on a handkerchief'; 'the work was written in a heavy hand on half a sheet of Baghdād paper'[214]; and 'the book was written on paper of the format of half a Baghdād sheet'.[215,216] Sakhāwi (d. 1245) tells in his history of Egyptian judges about a set of 40 volumes consisting of papers of different origins: five volumes of the largest format were on Syrian half sheets; three were on Complete paper; the others were of Egyptian paper, and strange to say, 'Frankish' paper of quarter sheet size. This shows that European papermaking in the first half of the 13th century had reached such a level of quality that their papers, probably from Italy and France, were able to compete successfully with the local Egyptian product.

Anyone familiar with the Oriental delight in display will not be surprised that from early times they were desirous of giving an opulent appearance to correspondence between rulers, and also to the more important state papers such as orders of investiture and presentations. The solemnity of the occasion was emphasised by the size of the document and its calligraphic flourishes. The usual size of paper was not sufficient for these official requirements, and a larger format was sought. This was not available by ladling pulp – the usual method of papermaking – so larger sheets were made by pasting sheets of paper together in the manner used to make papyrus rolls.(xv)

By this means, the technique of pasting papyrus *selides* together was transferred intact to the craft of papermaking. The papers of the Rainer Collection contain some noteworthy examples. Like papyrus *selides*, the widths of the sheets of paper were stuck together to make a roll. From measurement of intact rolls we can see that these sheet-lengths, originally widths, correspond to the medium width of the average papyrus fortes. Of these, the *hieratica* was 20.33cm wide, the *Fanniana* 18.48cm and the *Amphiteatrica* 16.6cm. These dimensions were still the average in later Arab times and corresponding to this, in each of the 'joined' pieces of 10th century paper, such as Paper 7276, each *selis* is 17.75cm and three *selides* have been joined together, a fourth having been torn off. Paper 10,915 has a *selis* of 19cm. Obviously, this superficial resemblance of the paper roll to papyrus roll helped the previously mentioned Pharaoh paper, the oldest of the papers of Khurāsān, to acquire its name.

The individual sheets intended to be joined together were called *tabaq* (plural *atbāq* or *tibāq*), that is, 'a sheet that fits to another'.[217] The line of glue was called *wasl* (plural *awsāl*), connection. The roll made by joining these sheets was *darj* (plural *durūj*).[218] One reads in the *Dīwān al-inshā*: 'Darj means in the common tongue, paper made longer by a series of joins (*awsāl*)'. The finished rolls of paper were distributed through the paper trade. Listed on Paper 8201 are items of trade including '30 rolls of paper'. Paper was cut and written on without regard to the glue line, resulting in a slightly uneven surface in places; there are hundreds of examples in the Rainer Collection. Paper 3074 (10th century) has been cut in such a way that two glue lines occur in the middle of the written area. Paper 2971 from the 11th century is another example.

Obviously one could order rolls of any length from the factory. The length of the roll or sheet was then *ma'hūd*, 'agreed by contract'. Bayhaqī mentions in his biography in Persian of Sultān Mas'ūd of Ghazna, dated 421H (1030AD)[219], that a diploma was written in *Qarmati* script on three sheets of paper that had been joined together. The previously mentioned vizier and historian Rashīd al-Dīn wrote about the maps he intended to draw up[220]: 'Considering that the intended purpose of the maps would be better and more easily fulfilled if they were drawn to the largest possible size, we specified by contract that each sheet of the map was to be made by joining six pieces (*atbāq*) of Complete Baghdād paper'. The Egyptian Mamluk, Sultan Baybars sent a letter in 1262 to the Mongol Benke Khan; it was a roll of enormous length, 51 metres, made of 70 Baghdād sheets.[221]

While a roll of any length could be produced and there were no norms for its dimensions, sheets were counted, folded and sold in fixed quantities and sizes. The standard changed with time and place: an ordinary sheet was called *talah* in Syria but *farkha* in Egypt. In the latter, a quire was called *dast*, from the Persian meaning hand, giving the French expression *main de papier*; in Syria they used *kiffa*, the Arabic translation of *dast*. A 'bundle' or 'packet' of paper was a *rizma*, which gave the Italian *risma*, the Spanish *resma*, the French *rame*, the English *ream* and the German *rizz* (14th century) or *reiss*.[222,223,224] Twenty-five sheets of *farkh* (3584.7 sq. cm) made one quire and five quires made one ream.

A 10th-century paper in the Rainer Collection (No. 8201) mentions 'two reams of *farkh* sheets'; and No. 8200 mentions various types of paper and also 'a quarter of a ream of paper at a price of one and two thirds of a carat'. Price depended not only on size but also on thickness and other qualities such as colour. There are many differing types of paper in our collection with dates ranging over some hundreds of years. Some of the finest Egyptian papers are thin and light, almost transparent; others are heavy but still smooth, strong, dense, opaque and parchment-like; again many others are rough and felt-like. Sometimes they are fine and felt-like or, as Paper 2215, woolly and hairy.

The following list shows some of the characteristics types:

1. Polished, thin, strong, parchment-like.
2. Rough, but fine, thin with a translucent body.
3. Felt-like, medium strong and opaque.
4. Felt-like, almost like cloth, dense and strong.
5. Smooth, very strong, parchment-like, opaque.

In addition to these differences, the paper came in a range of colours.

The following question arises concerning the colouring of the papers: what was the colour of the earliest papers and were they able to produce truly white paper at the beginning?

Our paleographers considered the question of colour; that is, they spoke of 'dyed papers', but they failed to tell us what they meant by 'undyed papers'. It does not so much matter whether the original colour of the earliest paper was white because we know that colour is more dependent on the method of preparation than on the material from which the paper is made. It depends on the processing of the rags by boiling, milling and bleaching.

Wiesner's research into this matter has produced important results: the colour of our papers varies greatly, leaving aside those papers which were dyed, and this variation can be ascribed to the process of ageing. The better quality paper must have been white or near white originally.

This statement has recently been checked against a large number of the earlier papers in the Rainer Collection, and Arab sources confirm it. Without doubt the Arabs knew then, as now, how to bleach either the unprocessed rags or the resultant paper pulp and so were able to make pure white paper. We do not know if the method of bleaching rags was any different from the treatment of raw linen, or how the paper pulp was bleached. It is clear that the sizing of the paper pulp with starch paste – and even more particularly, the application of starch powder to the paper surface – must have increased the whiteness considerably.

Many of our papers retain a striking whiteness today, due to the favourable conditions in which they were found, particularly where the surface was sufficiently protected. Therefore, our sources are correct.

The term *qarṭās*, a sheet of paper, also carries the implication of whiteness.[225] A white horse was described as '*qarṭāsī*, white as paper'.[226] Sheets of paper were also referred to as *durūj bayāḍ*, white sheets. The famous calligrapher, Ibn al-Bawwāb (d. 1022), wrote a poem on the subject of calligraphy and the following quotation is from the verses describing the making of ink: 'When this mixture has fermented long enough, take the pure white, smooth paper you have selected.'[227] The technical term *bayāḍ*, 'whiteness, white mark', denotes lacunae in old manuscripts and also refers back to the original colour of the writing material. Similarly the expression *'attaqa*, to make or let grow old, i.e. to make brown, in connection with paper, denotes a method of forging the age of a document.[228]

The art of making pure, white paper was spread by the Arabs to all countries reached by their language, including Europe by way of Spain.[229] The geographer al-Idrīsī wrote in 1176 about Bocayrente, and described one of their textiles as being of outstanding quality 'so that from the standpoint of whiteness, one cannot detect any difference from paper'. The pure whiteness of Arab paper even inspired some poetic effusions. Here is an example: when Louis IX was captured by the Saracens near Damietta, Sultan Tūrān-Shāh sent the French king's coat and a letter of victory to Amīr Jamāl al-Dīn, the governor of Damascus. The arrival of the valuable trophy made of scarlet cloth lined with ermine was praised by a shaykh as follows:

The coat of the Frenchman came
As a tribute to our great Amir
Its colour was white as paper
But our swords dyed it with blood[230,231]

Concerning 'dyed paper', Wattenbach states in *Schriftwesen im Mittelalter* that there is evidence for only one colour, blue, for which there are only two references, and the earlier of these rests on an incorrect translation from the Arabic.

The letter from the Byzantine Emperor to the Spanish Caliph 'Abd al-Raḥmān, written in gold allegedly on 'blue paper', was in fact a letter on blue parchment.[232] The Spanish Arab historian Ibn al-'Idharī[233] specifically described the reception of the Greek envoys and went on to say: 'and they handed over a letter from their ruler, written in gold ink on hyacinth purple parchment. This letter had a gold seal weighing four mithgals. On one side was picture of the Redeemer, Praise be to Him, and on the other side were pictures of Constantine VII Porphyrogenitus', who sent the message to the Caliph during the regency of his son Romanus in 338H (949AD).[234] The term used by Ibn al-Idharī and al-Maqqarī for the dyeing of parchment is 'dyeing it the colour of the sky', which corresponds to the Persian *āsmāngūn*, i.e. sky or hyacinth coloured, meaning purple, amethyst or hyacinth. This most beautiful and expensive of the colours was derived from a species of mollusc, *Purpura blatta*, and was used at the Byzantine court and also for royal robes of the Sāssānian Persians and the Arabs.[235]

The Emperor's blue letter to the Caliph would have been seen by many Muslims as offensive because blue was a colour held in contempt. It was the colour of the common headdress of the Christians as well as the colour of mourning for a great part of the Islamic world, although for the Umayyads of Spain, the colour of mourning was white. There is proof, however, that blue was used by the Arabs and the Persians for colouring paper. Without doubt in earlier times, this colour was only for monochrome papers. Dyes and plant juices were used either on their own or in mixtures to colour paper. Paper was dyed indigo or cobalt blue only for very specific purposes.[xvi] As I have mentioned before, *azraq* (blue) was used as much as *aswad* (black) by the Abbasi to signify mourning.[236] The death of a Persian ruler was marked by covering the minarets and pulpits with blue felt covers. Death sentences in Egypt and Syria were written on blue paper.[237] Blue was also the colour of renunciation in the sense of being true to God. In Persia, blue paper was used for wrapping medicines as it was considered more propitious than white paper. For the same reason, the followers of Shaykh Hasan Azraqpūsh, that is, 'the One dressed in blue', wore blue robes to show that their souls were dedicated to God and were free from earthly desires. They were the 'Observers' as opposed to another order of Sufis known as the 'Rose-coloured' or 'the Enjoyers'.

Red in all its shades was a very distinguished colour denoting joy and festivity.[238,239,240] Very popular (although the Prophet referred to it as Satan's colour[241]), the lighter shades such as rose-colour paper are often found in books. There are various examples in the Rainer Collection, and one 10th-century paper lists two types of rose-coloured paper. Darker shades of red such as cinnabar and ochre are found.[242] Red paper was considered the privilege of those of high rank, and a proof of special favour granted by the ruler to some of those in authority. Only the Regent of Damascus and the Governor of the Castle of al-Karak were allowed to use it for correspondence with the Court at Cairo.

But red was not only the colour of the 'Enjoyers'; it also became the colour of the populace. The oppressed and the poor wore this most striking of colours to attract attention to themselves. Shaikh Abū Sa'īd of Baghdād gave a sermon rebuking Nizām al-Mulk (d. 1092AH), the famous Seljuq vizier. In it, he told of a deaf Indian king who is said to have ordered all the poor and oppressed to wear red garments. No one else was allowed to dress this way. Thus, he could immediately recognise the condition of a person without the trouble of asking.[243] This is a likely story for it was a common habit throughout Persia that those asking a favour from the king, or making a complaint, would wear red paper – they just stood in his path in order to attract his attention. In the law courts, the defendant did the same before the judge. Vassāf tells of the 'paper shirt of the oppressed'; the Persian expressions 'a shirt from paper', 'paper garment' and 'a shirt of paper' are synonymous for the defendant.[244] Hāfiz says:

I must weep bloody tears
Onto the beggar's paper garment
Because for the humiliated
There is no comfort of justice

Yellow was the next most popular colour after red. The Arabs followed the Persian example for dyeing their writing materials yellow from the 7th and 8th centuries onwards. Al-Balādhurī (d. 892) explains in some detail that before the introduction of paper, both parchment and papyrus had been coloured with saffron.[245] This colour had a great reputation in all crafts. The description of the 'saffron-coloured cheeks' of passionate lovers whose melancholy was painted on their cheeks, shows what yellow-coloured paper sometimes signified to the man-of-letters: the 'gold of a lover', says Jalāl al-Dīn Rūmī, 'is a yellow face'.[246] As far as I have been able to ascertain, there is only one strikingly yellow piece among our papers, although others do show a certain yellowish tint. Everyone familiar with Oriental manuscripts must be impressed with the variety of shades produced by Arab and Persian paper dyers. Besides pale yellow, we also find deep yellow, pea yellow, bright orange papers and *papier chamois*. The use of grey, green and pale blue for book papers is rarer. A study in depth of the uses of different kinds of coloured papers, particularly speckled papers, belongs to the field of palaeography and would take us too far from our subject.

12

On Arab diplomacy

With these words I come to my final subject connected with Arab papermaking. I want to remark straight away that I am touching on an area of critical examination which has been more neglected than any other field of Oriental studies. In general, we know nothing about Arab documents. There should be an exhaustive search of the thousands of historical documents concerned with diplomacy, starting with the Rainer Collection; however it is not my intention to start such a task here and now. Let me recall some dates of diplomatic importance which relate to our Egyptian papers. Earlier, I tried to pinpoint when paper was first used in Arab government; this important event is presumed to have taken place between 794 and 795AD. Until then, parchment and papyrus were used; one or the other was favoured depending on the conditions prevailing at the time or place. Mu'āwiyya, the first of the Umayyads, chose parchment as the main writing material when he acceded to the Caliphate. By this means, he hoped to give a special importance to the edicts issued from his office.[247] His successors up to the end of the dynasty usually used papyrus.[248] There was a return to parchment under the 'Abbāsi, particularly under Abū al-Abbās al-Saffāh and his first vizier, Khālid ibn Barmak. This same Barmaki had held under the Abbasi al-Mansūr, the position of Governor of Fārs (Persia) as well as the offices of war and of finance for that region. He had the registers made up in book form using parchment and papyrus, instead of the scroll form used until then.[249] Alternation between parchment and papyrus continued until the time of Hārūn al-Rashīd by which time paper had become a common import from Khurāsān.[250] It was then that the change to paper took place in the government offices of Baghdād.

By examining documents recorded in registers, particularly those concerning paper, and by looking at the bookkeeping of the government offices in Egypt and

Syria we can see that rules governed the choice of paper for particular functions.[251]

We can divide the types of paper used as follows:

1. Baghdād paper

It was introduced into Egypt but was always rare. In government offices, it was used exclusively for contracts, orders of investiture and edicts from the ruler whose entire correspondence was written on this paper. However, the fierce competition from a flourishing paper industry in Damascus led to a diminution in the use of Baghdād paper in Egyptian offices.

II. The so-called 'Syrian paper' made in Damascus

There were three different types:

a. Hama paper, named after the town where it was originally made. Production was transferred to Damascus, but the method of production remained the same. Hama paper was used only occasionally in Egyptian government offices after the 14th century.
b. Syrian paper, or more accurately Damascus paper – *charta Damascena* to the West. This most famous of Syrian papers left no room for competitors in Syria itself. I have listed previously the names used for sheets and reams of this type of paper. It was used in Syria, the Eastern provinces, Yemen, Rūm (Asia Minor) and the Hijaz, and also for various official purposes in Egypt. In the offices of the court, it was employed for orders and similar documents and in the dispatch departments, for registers. This Syrian paper was also used for correspondence and commissions in the exceptional circumstances of the clerks running short of Egyptian paper while accompanying the ruler on his travels. In such circumstances, the secretary for secret correspondence needed written permission from the Sultan. Egyptian paper was so highly praised in all regions and at all courts that Syrian paper was substituted only in the more urgent cases. As mentioned previously, the Syrian Regent and the Governor of al-Karak were both entitled to use red Syrian paper for correspondence with the Court in Cairo.
c. Bird paper, also called Dispatch paper. Extremely thin and weighing only 1½ drachms per sheet, it had two uses: for secret letters of amorous intent, and for correspondence by carrier pigeon. In this latter case, the dispatches were tied to the wings of birds, as I will describe later.

III. Egyptian paper

One must distinguish between the following types:

a. Mansūr paper, Complete, or Perfect paper (also known as al-Kāmil), probably so called after the Fatimid Caliph Abū 'Alī al-Mansūr al-Amir bi-ahkām Allah, 495–524H (1102–1130AH). It was very strong with a width of one ell (48.886cm), length of 73.329cm, area 3584.7 sq. cm. It was used primarily for diplomas for the investiture of the governors and for the bulk of correspondence from the dispatch bureau. Muhammad ibn 'Umar al-Madīnī (1108–1185) says that under the Caliph of Baghdād, the paper was available in five different sizes, specified by law. In his *Book of the Pen* he records:

Two-thirds sheet, 32.5cm wide by 48.8cm long for a letter to the Caliph.

Half sheet, 24.4cm wide by 36.6cm long for the amirs.

One-third sheet, 16.2cm wide by 24.4cm long for the superintendants and secretaries.

Quarter sheet, 12.2cm wide by 18.3cm long for merchants and others of this class.

One-sixth sheet, 8.1cm wide by 12.2cm long for mathematicians and surveyors.

These specified uses for this paper remained customary in Egypt with only slight changes into the 15th century. A second smaller Mansūr paper (44.814 x 67.221cm) also came into circulation.

b. Nine different types of paper were used in the dispatch offices.

1. Complete *Tūmār*, first made in 815H (1412AH) at the start of the reign of the Circassian Mamluk Sultan al-Malik al-Mu'ayyad Shaykh. Similar in style to Complete Baghdād paper, it was 1½ ells wide (73.329cm) by 109.9935cm long and with an area of 8065.7 sq. cm.[(xvii)] This size could just be made efficiently by the ladling method. Imām al-Musta'īn bi-Allah used this paper for the document of investiture of al-Mu'ayyad Shaikh. This was in 1412 when al-Musta'īn bi-Allah, like the earliest 'Abbassi, reigned as both Caliph and Sultan. In that year, all power outside the Caliph's sphere was transferred to the Shaikh who continued as Sultan. From then onwards, all similar documents were issued on Complete *Tūmār*.
2. Baghdād paper of Egyptian manufacture was so called because its size was the same as the Baghdād half sheet, one ell, 48.886cm wide by 73.329cm long, with

an area of 3584.7 sq. cm. From the start of the Mamluk dynasty, the Egyptian sultans had contracts written on this paper; also they used it for letters to the Khans and the rulers of Īran and Tūrān. When a shortage of this paper or the original Baghdād half sheet occurred, the slightly smaller Mansūr was used, as mentioned previously. It was on just such an occasion that the second type of Mansūr paper was used for a letter to Muhammad Khān, ruler of Khwārizm and Dasht-i Qipchāq in 832H (1429AH).

3. The smaller Baghdād paper of Egyptian manufacture. A sheet measuring four fingers in width less than Complete Baghdād, that is 65.185cm wide, 97.7775cm long and with a width of 6360.2 sq. cm. It appeared during the dynasty of Circassian Mamluks and was used for the document of investiture of the first ascent to the throne of Sultān al-Malik al-Nāsir Faraj (1399). The appropriate Baghdād paper was not available, and his successors continued to use this paper for the same purpose.
4. The 'two-thirds of a *qat'* ' paper. So called because it was two-thirds of the size of Complete Mansūr, ⅔ of an ell, 32.5906cm wide by 48.886cm long with an area of 1586 sq. cm. It was the paper used for the investiture of Grand Governors, appointments of governmental heads, the ministers, majordomos, inspector of the army, the four High Court judges, and the Governor of Alexandria.
5. Mansūr half sheet. As its name implies it was half the size of Complete Mansūr, ½ ell wide (24.443cm), 36.6645cm long and had an area of 893 sq. cm. It was used for some orders of investiture; for most of the imperial letters of clemency; for the commissions of the officers of the Tablakhāna; for the officers commanding a thousand men in the garrisons of Syria; for letters to rulers of the Middle states (the second class kings); and finally to those of secular and clerical eminence, the highest Shaikh and the highest secular and the clerical civil servants of Syria, especially in Damascus and Aleppo.
6. Mansūr one-third paper has the size of a third of Complete Mansūr. It is ⅓ ell wide (16.2953cm), 24.443cm long, and with an area of 395.2 sq. cm. It was used for most communications, for imperial pardons for persons of lower rank, for minor instructions to commanders of the forts, rulers of the third rank, and to officers of the same rank as a Persian Amiri Ulus. It was also used for letters of patent to the Amirs of Horsemen, as well as to all clerical officers of third rank.

   The size of this much-used ⅓ Mansūr paper was strictly regulated; its use was dictated by actual rank, not by mere title. The procedures for issuing documents to members of secular and clerical hierarchy was regulated to an extent that allowed no deviation by the secretarial staff. It would never have occurred to a scribe to cut a ⅓ Mansūr, or to stick two pieces together to make a larger sheet of paper.
7. Common paper with a width of a quarter of an ell plus one carat, that is 14.257cm wide by 21.38625cm long and with an area of 302.4 sq. cm. It was

used for certain decrees, short dispatches, notices of employment for officers and men of the Halqah troop, for the Turcomans who fought against the Infidel, for announcements of amnesties, for the registering of oaths, for maps of the routes of marches and for general correspondence except that to certain rulers. The so-called Common paper was required in enormous quantities, being the most often used in administrative offices.

8. Secret dispatch paper ('bird paper' was usually used). The format was strictly regulated, but was left to the head of the Secret Service to decide what was required under the circumstances.
9. Dispatch paper (used for carrier pigeons). Made in Egypt, it was lightest and smallest size of paper, similar to that made by the Damascus mills. Three fingers wide (6.108cm) and 9.161cm long, with an area of 55.95 sq. cm. The size did not vary regardless of whether it was used for imperial dispatches or any other. This thinnest of writing materials was made specially for the dispatch department and had the official name 'bird paper'. The cost of this paper was paid for by the profits from the workshops in Cairo that dyed silk.

   A 'slip of paper' or a 'ticket' was tied to the wing of the carrier pigeon (usually the Blueback type). The pigeon stations were separated by a distance of three ordinary post stations. The dispatch was removed from the feathered messenger and tied to the wings of the next bird. This continued from station to station until the last pigeon arrived at the Sultan's palace on the hilltop citadel of Cairo. The warden of the pigeon loft took the pigeon to the head of the Secret Service who would remove the dispatch and read it. In this way news arrived daily in the capital, from Egypt and from Syria informing the ruler of murders, robberies and fires.[252] Again one sees that the consumption of the thinnest and most valuable paper was enormous.

So much for what history tells of the types of paper used in Egypt. I will have to save a study of the significance of the documents in the Rainer Collection for a later work. This section is completed with a table summarising the sizes of papers I have described.

Paper sizes used in the Egyptian government offices during the Middle Ages

	Width in			Width	Length	Area
	Ells	Fingers	Carats	(cm)	(cm)	(sq. cm)
Made in Baghdad						
I. Whole sheets	1½			73.3	109.9	8065.7
II. Half sheets				48.8	73.3	3584.7
Made in Egypt						
I. Complete *Tūmār*	1½			73.3	109.9	8065.7
a. Baghdad	1	–4		48.8	73.3	3584.7
b. Baghdad incomplete	1½			65.1	97.7	6360.2
c. ⅔ *qat'*	⅔			32.5	48.8	1586
d. Half Mansur	½			24.4	36.6	893
e. 1/3 Mansur	⅓			16.2	24.4	395.2
f. Common paper	¼		+1	14.2	21.3	302.4
g. Bird paper		3		6.1	9.1	55.9
II. Complete Mansur	1			48.8	73.3	3584.7
⅔ Complete Mansur	⅔			32.5.	48.8	1586
½ Complete Mansur	½			24.4	36.6	893
⅓ Complete Mansur	⅓			16.2	24.4	395.2
¼ Complete Mansur	¼			12.2	18.3	223.2
⅙ Complete Mansur	⅙			8.1	12.2	98.8
III. Smaller Mansur	1	–2		44.8	67.2	3012.4

13

Conclusions

The first result of my small-scale study is the beginning of a history of Arab paper (if this claim is not too immodest) from its first appearance up to the end of the Middle Ages. It can be summed up by the following three points:

1. It is no longer possible to support the idea that the origin of paper is wrapped in clouds that can never be swept away, or that it is futile to try to connect its invention with a particular person or period.

   The history of paper is illuminated from new sources; its main events have been reassessed, particularly its chronology. Previously, Arab papermaking and paper distribution have been placed too early in time, our knowledge having been based on incomplete information and misinterpretation of the sources. Papermaking in Islam does not date from 650, 676 or as assumed by most people, 704AH. We now can state with certainty that the year 751AH was the beginning of papermaking in Arab lands. Paper was little known in the Western Christian world until after a second state papermill had been built in Baghdād in 794–95 from which the Arabs distributed the new writing material to the West.
2. When papermaking first started in the Orient, the new writing material had to compete with the papyrus which held the market, particularly in the West. The much-ventilated but until now unresolved question of the date of the final decline of papyrus as a result of the supremacy of paper can at last be answered satisfactorily. Dated examples of both writing materials exist in abundance in the Rainer Collection, and this information has been combined with Arab sources. The decline of Egyptian papyrus making was believed to

have taken place in the 12th century, but our sources show that it can be set back to the second half of the 10th century. The new papyrus industry in Sicily was connected with this demise; earlier statements describing a papyrus industry on the island dating back to the 8th century have now been proved wrong. The evidence for its early production was the famous Bull of Pope John VIII dated 876AH on Sicilian papyrus. This scroll has now been shown to be made of Egyptian papyrus.

3. The controversy about whether cotton or linen is the older paper fibre is now irrelevant. The earliest sources on the question of the origin of paper are Arabic, and they do not mention the existence of cotton paper. The first papers in the region were made from linen rags. Therefore, the theory that linen paper developed from cotton paper is untenable. This is confirmed by the microscopic study of the earliest papers in the Rainer Collection, some of which date back to the time of the early Arab papermaking. The historic texts provide information about hemp paper, and about sizing and filling paper with wheat starch. Study of the papers in the Rainer Collection shows that the Arabs knew how to make laid paper by ladling pulp onto a wire mould from the very start. The story that from the time of its introduction to the Arabs up to the 13th century paper was made from raw cotton fibre can probably be ascribed to a confusion of names brought about by the surface characteristics of the paper.

As any expert in the field will admit, very important facts about the development of this writing material have emerged. These will diminish the pride of discovery that was claimed quite wrongly by our time and our part of the world. The history of Arab paper and the cultural movement within the Islamic world makes us recognise the full revolutionary impact of paper on this huge region.

A.v. Kremer says:

> From a cultural and historical point of view the reduction in the cost of writing material, which went hand in hand with the production of paper, was of great importance. Books on parchment or papyrus were so expensive that they were available to very few. By the production of a cheap writing material, and its supply to markets both east and west, the Arabs made learning accessible to all. It ceased to be the privilege of only one class, initiating that blossoming of mental activity which burst the chains of fanaticism, superstition and despotism. So started a new era of civilisation. The one we live in now
>
> (*Culturgeschichte*, II, 308)

Author's notes

1 *Microskopische Untersuchungen der Papiere von el-Faijum*, 1, 45ff.

2 *Marāsid al-ittilā'*, I, 69.

3 *Abū al-Fidā, Taqwīm al-buldān*, Schier (ed.), 91.

4 *Al-Nawawī, Kitāb tahdhīb al-asmā*, Wüstenfeld (ed.), 710, 712. Ibn Khallikān, *Kitāb wafāiāt al-a 'yān*, Wüstenfeld (ed.), 271. Ibn Abi al-Mahāsin, *Annals*, Juynboll (ed.), II, 79,234.

5 Prof. de Goeje gives reference to the Qarmatians in the second edition of *Memoire sur les Carmathes du Bahrain et les Fatimides*, 1886, 199ff. Ibn Durayd, *Jamharat al-lugha*, III, fol. 391; al-Jawhari, *al-Sihah*; Ibn Khallikān, No. 186, page 124.

6 Al-Suyūti, *Husn al-muhādara*, Bulaq edition, 1299H, II, 230.

7 See *Mitteilungen*, I, 96.

8 *Kitāb al-buldān*, Juynboll (ed.), 39.

9 Craftsmen were brought from all parts: workers in glass and ceramics from Basra, also the makers of mats known as *Husur*; from Kufa came skilled potters and perfumers.

10 Al-Tha'ālibī (d. 1038AH). *Latāif al-ma'ārif*, de Jong (ed.), 97. Al-Suyūti, l.c.II, 238.

11 *Mukhtasar kitāb al-buldān*, de Goeje (ed.), 66

12 Al-Tabarī, *Annals*, S. Guyard (ed.), III, iv, 999.

13 Ibn al-Athīr, *Chronicles*, Tornberg (ed.), VI, 377.

14 Al-Qazwīnī, *'Ajā'ib al-makhluqāt*, Wustenfeld (ed.), II, 214.

15 Ibn Abi al-Mahāsin, *Annals*, Juynboll (ed.), II, 208

16 Al-Muqaddasī, *Kitāb ahsan al-taqāsīm*, de Goeje (ed.), 239, 'All the Spanish Arabs' copies of the Quran and their books are written on parchment', but this obviously refers only to special types of manuscript. Among the 'books', i.e. the secular manuscripts, there must have been quite a number of exceptions.

17 Ibn 'Abd Rabbihī, *al-'Iqd al-farīd*. Bulak edition, 1293, 223. The Spanish word *albardin* is derived from the Arabic word, *al-bardī*, the papyrus plant; in the dialect of Valencia, *albardi* is used unchanged.

18 *Al-'Iqd al-farīd.*

19 *Kitāb ahsan al-taqāsīm*, de Goeje (ed.), 32ff, 193ff, 202ff.

20 Muhammad ibn Ishāq, *Kitāb al-fihrist*, Flugel (ed.), I, 21.

21 The following passage from the *Fihrist*, continues on from the previous quotation, but refers to a much earlier time: 'The Byzantines write on white silk, parchment and other materials, on Egyptian papyrus and on *al-faljān*, that is parchment from the skin of a wild donkey.' About donkey parchment see Wattenbach (*Schriftwesen*, 98). The word *al-faljān*, 'the two halves', he uses to mean Duplices or Diptycha, the writing surfaces of which were sometimes laid with parchment instead of wax. The use of *al-faljān* in connection with parchment is repeated in the *Fihrist* as *jalūd faljān* or *faljān* parchment (see pp. 40, 353, and II, 188). The custom of writing only on the inside of a folded sheet was continued by the Arabs in later times, as shown by a 13th-century example in the Rainer Collection.

22 *Kitāb al-masālik wa al-mamālik*, de Goeje (ed.), 85f.

23 Al-Maqrīzī, I, 186.

24 Wattenbach, *Das Schriftwesen im Mittelalter*, 1875, 85. Gardthausen, *Griechische Paläographie*, 35.

25 Ibn Baytār, Arabic text, I, 86ff. The term *qartās* was used by al-Maqrīzī, II, 274 and al-Qalqashandī (ed. Wustenfeld, 198) and could be taken to mean papyrus sheet, and thus became evidence for the use of this writing material in later Fatimid times. It means, however, a pomander filled with sweet-smelling spices. Dozy, suppl.

26 II, 321 L.c. and Heyd, L.c 88 and Heyd: *Geschichte des Levantehandels*, I, 99.

27 Tardif, *Archives de l'Empire; chartes et diplomes*, Paris, 1864, 74. Gardthausen, 1.C.35

28 *Kitāb al-masālik wa al-mamālik*, 85f.

29 *Storia dei Musulmani di Sicilia*, II, 1858, 299.

30 *Charte latine sur Papyrus d'Egypte de l'année 876*, Paris, 1835, pl. I.

31 *Oesterrechische Monatschrift für den Orient*, 1885,164.

32 About the Egyptian Director of Finance and his officers, see *Mitteilungen*, I, 5ff, 299.

33 Amari, *Storia dei Musulmani di Sicilia*, II, 299. Paoli, Cesare, *Del Papiro*, Florence, 1878, 37f.

34 *Memoires de l'Academie de Sciences*, Paris, 1854, XII, 469ff. There is no historical evidence for the claim in the *Memoire* that the Syrian species of papyrus was imported into Sicily by the Arabs. Paoli Cesare, l.c.37 and Th.Birt, *Das antike Buchwesen*, 223, agree about this.

35 A similar event took place when, in Justinian's time, many silk workers from Taurus and Berytus moved to Persia as a consequence of a monopoly that was damaging private enterprise. Procopius, *Anecdotes*, Bonn (ed.), III, 140ff.

36 I must add here that Professor Wiesner, as he says himself, approached the subject without pre-knowledge; in other words, he was without preconceptions. He knew nothing of my historical research which was to confirm his microscopic examination; our studies developed completely independently of each other.

37 *Das Schriftwesen im Mittelalter*, 115.

38 St. Julien, *Industries anciennes et modernes de l'Empire Chinois d'après des notices traduites du chinois*. Paris, 1869, 145.

39 *ZDMG* (*Zeitschrift der Deutschen Morgenländischen Gesellschaft*), I, 224.

40 Keferstein in *Allgemeines Encyklopädie* by Ersch and Gruber, 3, XI, 84.

41 'Alī ibn Muhammad al-Fārisī in *Burhān-i Qāti'*, Calcutta, 1818.

42 Reinaud, *Mémoire géographique, historique et scientifique sur l'Inde*, 1849, 305.

43 Casiri, *Bibl, Arab. Hisp.*, II, 9.

44 The year 88H begins 12th December 706 and ends 30th November 707AH.

45 See Schafer, *Geschichte von Spanien*, II, 126.

46 Muhammad Husain ibn Khalif al-Tabrīzī, *Burhān-i Qāti'*, Calcutta, 1818.

47 *Archivio storico italiano*, 1885, XV, 230.

48 Al-Tabarī, *Annals*, Guidi (ed.), II, iv, 196. Al-Fāsī, *Shifā' al-gharām*, Wustenfeld (ed.).

49 *Kitāb al-fihrist.*

50 Ibn al-Athīr, *Chronicles*, Tornberg (ed.), IV, 403.

51 Al Bāladhuri (d. 893AH), *Kitāb futūh al-buldān*, de Goeje (ed.), 415. Al-Tabarī, *Annals*, Thorbecke

(ed.), II, i, 179. *Ta'rīkh Abī al Fidā'* Constantinople edition, 1286H, I, 97.

52 Al-Ya'qūbī, *Kitāb al-buldān*, Juynboll (ed.), 74.

53 Al-Bāladhuri, l.c.421f. Al-Tabarī, l.c.II, iv, 124lff, 1252.

54 *ZDMG*, VIII, 529.

55 This quotation provides unambiguous evidence for the date of the decline of Egyptian papyrus production.

56 *Latāif al-ma'ārif*, de Jong (ed.), 126.

57 *'Ajā'ib al-Makhlūqāt*, Wüstenfeld (ed.), II, 360.

58 Ibn al-Athīr, *Chronicles*, Thornberg (ed.), V, 344. *Ta'rikh Ibn Khaldūn*, Bulaq edition, 1284H, II, 178.

59 Ibn Hawqal, 390. Al-Muqaddasī, l.c., 263

60 *Kitāb al-fihrist*, l.c.21.

61 *Kitāb al-fihrist.*

62 Herodotus, VI, 20.

63 Al-Mas'ūdi, *Murūj al-dhahab*, Bulaq edition, I, 124.

64 Nicetas, Bonn edition, Lib.II, l, 99; II, 8, 129f. Otto Frising, *Epist. de gestis Friderici I*, Lib. I, XXIII, in *Monum. Germ. Hist. Script*, XX, 370.

65 Al-Muqaddasī, l.c., 148.

66 *Kitāb al-fihrist*, l.c., 21. *Dīwān al-inshā* in Rashīd al-Dīn, *Hist. des Mogols*, Quatremère (ed.), CXXXIV.

67 Ersch and Gruber, l.c., 105. St. Julien, l.c., 145.

68 Ibn al-Faqīh, l.c., 316

69 'Izz al-Dīn 'Abd al-Azīz al-Qasīm, *Kitāb al-mukhtār min nuzhat al-nāzir* by Casiri, l.c., I, 209. *Dīwān al-inshā*, CXXXIV.

70 Ibn al-Faqīh, l.c., 21.

71 *Kitāb al-fihrist*, l.c., 21

72 De Sacy, *Mémoires de l'Academie des Inscription* etc., I, 188f. Sprenger, *Das Leben und die Lehre des Muhammad*, III, xcii. Gardthausen, *Griechische Paläographie*, 49.

73 Literally … 'that the cotton for Khurāsān and the flax for Egypt'. Al-Tha'ālibī, *Latāif* etc., 97. *ZDMG*, VIII, 526, where 'linen' is written for 'hemp'.

74 'Alī al-Jawharī, *al-Durr al-thamīn*, in the Royal Library in Vienna, fol. 71v, in which *kattān* is vocalised as *kittān* following the Egyptian dialect. The Rainer Collection has a number of these writing linens with Coptic and Arabic texts.

75 G. M. S. Fischer in Ersch and Gruber's *Encyclopädie*, l.c., 90.

76 G. Ch. Lichtenberg.

77 *Mustalahāti bahāri-i 'ajam*, Calcutta, 1853.

78 *Kitāb al-aqālīm*, de Goeje (ed.), 288.

79 *Kitāb al-masālik wa al-mamālik*, l.c., 337.

80 *Kitāb ahsan al-taqāsim*, l.c., 326.

81 Gardthausen, *Griechische Paläographie*, 48. He was wrong, therefore, when he reports that al-Muqaddasī does not list paper among the exports of Samarqand.

82 *Kitāb al-fihrist*, l.c., 21

83 Nāsir-i Khusraw, *Safar nameh*, Schefer (ed.), 41.

84 Al-Tha'ālibī in *ZDMG*, VII, 526.

85 Al-Qazwīnī, *'Ajā'ib al-makhlūqāt*, l.c., II, 360. Al-Suyūti, *Husn al-muhādara*, Bulaq edition, 1299H, II, 288. Hājjī Khalifa, *Jihān Numā*, Constantinople, 1145, 350.

86 *Mustalahāt-i bahār-i 'ajam*, Calcutta, 1835.

87 Poncelin de la Roche Tilhac, *Philosophische Beschreibung des Handels und Besitzes der Europaer in Asien und Africa*, I, 25.

88 The name Semiscant used in the West in the Middle Ages is strikingly similar.

89 Bretschneider, *Notes* etc. 1875, 38, note 93; 45.

90 A.v. Kremer, *Culturgeschichte des Orients*, II, 307f.

91 *Muqaddimah*, Bulaq edition, 1284, I, 352. *Prolégomènes*, Arabic text by M. Quatremere, I, ii, 350.

92 Which is why they could afford expensive parchment.
93 *Khitat*, Bulaq edition, I, 91.
94 *Dīwān al-inshā*, in Rashīd al-Dīn, *Hist. of the Mongols*, Quatremere (ed.), I, CXXXIV.
95 He was appointed in 177H (793AD) but did not arrive in Khurāsān until the following year. Ibn al-Athīr, l.c.VI, 96, 100, 101.
96 Ibn al-Athīr, l.c.VI, 104.
97 *History of Egypt under Sultan Nasr from 691–741H*, Codex 406 in the Royal Library, Munich, fol. 59v. Al-'Asqalani, *Al-durar al-kāmina*, Arabic ms. in the Royal Library, Vienna.
98 *ZDMG*, XVIII, 399.
99 Yāqūt, *Mu'jam al-buldān*, Wüstenfeld (ed.), II, 522.
100 *Kitāb al-fihrist*, l.c.40.
101 Al-Hamdānī, *Kitāb sifat jazīrat al-'Arab*, D. H. Muller (ed.), 55.
102 Al-Muqaddasī, l.c. 100.
103 The Papyrus Find at al-Fayyūm in *Denkschiften der kaiserl. Akademie der Wissenschaften*, Philos.-histor.Classe, XXXIII, 14.
104 Al-Maqrīzī, I, 102, where the times of sowing and ripening have been confused.
105 Al-Nabrāwī, *Nihāyat al-rutba fī talab al-hisba*; ms. in the Royal Library, Vienna, N.F.272, fol. 28v.
106 I quote from papers 4110, 7164 and 7246. I might add that, in Arab times, the Egyptians cultivated cotton (Gossypium), see papers 368 and 9583 to 9585. There is evidence from al-Maqrīzī (d. 1442) for the presence of cotton in Nubia. In Egypt in those days, cotton was sown in Pharmuthi, 27th March–27th April, 4 mass (approximately 5 quarts) of seed to the feddan. It was ripe in Thoth, 29th August to 27th September, giving a yield of 8 centner (400kg) per feddan. Paper 7281 says that 'Abd Allāh ibn al-Walīd, a cotton planter, paid 3 2/3 dinars for outstanding tax in the Ashmūn Administrative districts'.
107 *Die Persische Nadelmalerie Susandschird*, Leipzig, 1881, 187f.
108 Ibn Abī al-Mahāsin, *Annals*, Juynboll (ed.), II, 8.
109 *ZDMG*, VIII, 214.
110 *Safar nāmeh*, Ch. Schefer (ed.), Persian text, 53.
111 'Abd al-Latīf, *Relation de l'Egypte*, S. de Sacy, 198.
112 Al-Ya'qūbī, *Kitāb al-buldān*, l.c. 126f. A.v. Kremer, *Culturgeschichte der Chalifen*, II, 305.
113 Al-Maqrīzī, II, 10, 23. The name *al-warāqīn* does not mean 'the bookseller', for whom al-Maqrīzī, II, 102, uses another expression when he writes about *sūq al-Kutubīyīn*, booksellers' market.
114 Al-Muqaddasī, l.c.159.
115 Al-Muqaddasī, l.c.181.
116 Ibn Battūtah, *Kitāb al-rihla*, Bulaq edition, 1287H, I, 53.
117 Al-Muqaddasī, l.c. 180
118 Al-Muqaddasī, l.c. 162
119 *Safar nāmeh*, Ch. Schefer (ed.), 41.
120 *Dīwān al-inshā.*
121 Al-Bakrī, *Kitāb al-maghlib*, de Slane (ed.), Alger, 1857, 115. *Kitāb al-istibsār fī 'ajā'ib al-amsār*, A. v. Kremer (ed.), 69
122 See my *Persische Nadelmalerei Susandschird*, 124f.
123 Abū al-Hasan, *Al-qartās*, Tornberg (ed.), 35.
124 *Al-qartās*, l.c. 26.
125 *Al-'Iqd al-farīd*, l.c.II, 223.
126 Al-Maqqarī, *History of the Mahometan dynasties of Spain*, P. Gayangos, II, 139, 169.
127 Al-Idrīsī, *Description de l'Afrique et de l'Espagne*, Dozy and de Goeje (eds), 192 (text).
128 Al-Maqqarī. l.c. I, 67, 94.
129 *Mu'jam al-buldān*, Wustenfeld (ed.), III, 235.
130 Ersch and Gruber's *Allgemeine Encyclopedie*, l.c. 85, note 23.
131 *Griechische Paläographie*, 51.

132 Wattenbach, l.c. 118
133 *Hist. nat. lib.*19, cap.2.
134 For further references, see my essay *Ueber einige Benennungen mittelaltlicher Gewebe*, I, 1882. 2ff.
135 Ibn Jubair, *Rihla*, W.Wright (ed.), 227.
136 Yāqūt, *Mu'jam al-buldān*, I, 822. Al-Qazwīnī, *'Ajā'ib al-makhlūqāt*, II, 227. Barbier de Meynard, *Dict. geogr. de la Perse*, 133.
137 Ibn al-Athīr, *Chron.*, Xii, 328f.
138 *Archivio storico italiano*, 1885, XV, 230–34.
139 *ZDMG*, VIII, 214.
140 *Recueil de voyages et de mémoires*, publ. by La Société de Géographie, 384.
141 Oderico da Pordenone, *Peregrinatio*, Yule (ed.) (Cathay and the way thither), II. App. I, XXVI.
142 *Itinarium Willelmi de Rubruck*, D'Avezac (ed.) in *Recueil de voyages et de mémoires*, publ. by La Société de Géographie, IV, 329. Also publ. Guillaume de Rubruck, *Récit de son voyage*, transl. L.de Backer, Paris, 1877, 194.
143 Heyd, *Geschichte des Levantehandels*, II, 251.
144 Wattenbach, l.c. 116.
145 C. Paoli, l.c. 231
146 Strabo, XVI, 748. Pliny, *Hist.nat.*, V. 23. Plutarch Anton. c. 37. *ZDMG*, VIII. 209ff.
147 Leo Diak., IV. 10; X, 4.
148 R. Pococke's *Bezchreibung des Morgenlandes*, II. 242.
149 Malalas, *Chron.*, L, XIII. Ammianus, XIV. 8.
150 Ibn al-Athīr, *Chron.*, I, 238.
151 Al-Tabarī, *Geschichte der Sasaniden*, Th. Noldeke, 239.
151 Ibn al-Athīr, l.c.l, 317.
152 Ibn al-Athīr, *Chron.*, l.c.II, 386.
153 Leo Diak., S. 71.
154 Leo Diak., S. 166–8. Ibn al-Athīr, *Chron.*, VIII, 423.
155 Cedreni, Opp. II, S 673.
156 Ibn al-Athīr, l.c. X, 69.
157 L.c. 98, 322, 325, 338.
158 *Guill. Tyr. Hist. rerum in partibus transmarinis gestarum*, XXX, ll. Ibn al-Athīr, l.c. X, 436. Abū al-Faraj, *Ta'rīkh mukhtasar al-duwal*, Pocock (ed.), 379.
159 Ibn al-Athīr, l.c. X, 458.
160 Abū al-Faraj, l.c. 422, 426.
161 Abū al-Faraj, l.c. 532.
162 Nāsir-i Khusraw, *Safar nameh*, l.c. 31.
163 Ibn Jubair, *Rihla*, W. Wright (ed.), 250f.
164 *'Ajā'ib al-makhlūqāt*, Wüstenfeld (ed.), ii, 182.
165 *Beschreibung des Morgenlandes*, 1754, II, 242.
166 *Kitāb nukhabat al-dahr*, Mehren (ed.), 205f.
167 *Taqwīm al-buldān*, Schier (ed.), 152.
168 *Kitāb mu'jam musta'jam*, Wüstenfeld (ed.), II, 543.
169 Sathas, *Bibl. greaca medii aevi*, I, 66, 67, 50.
170 *Diwān al-inshā*, l.c. CXXXV.
171 Montfaucon, *Dissert. sur la plante app. papyrus*, in *Mem. de l'Acad. des inscr. T.*, IX, 326.
172 *Theophili Presbyteri Diversarum artium schedula* I, A. Ilg (ed.) in the *Quellenschriften für Kunstgeschichte*, VII, 1874, 51.
173 A. Ilg, in the introduction to his edition of Theophilus, l.c. XLI.
174 St. Julien, *Industries anciennes et modernes de l'empire chinois*, 141, 145, 149.
175 Wattenbach, l.c. 347.
176 Sathas, *Bibl. Graeca medii aevi*, I, 68.

177 As does Gardthausen, *Paläographie*, 49. See also Fischer in Ersch and Gruber's *Allgemeiner Encycl.*, 83.
178 Al-Maqrīzī, 'Rasā'il', in de Sacy's *Chresth. arabe*, II, 473.
179 *ZDMG*, VIII, 214.
180 Abū al-Fidā, *Kitā b taw qī m al-buldā n*, Schier (ed.), 152. Hājjī Khalīfa, *Jihān Numā*, Constantinople edition, 598.
181 *Lisān al-'arab*, IV, 384. Abū al-Hasan, *Al-qartā s*, Tornberg (ed.), 26, 35. *Inventaire de bien d'un juif marocain, nommé Musa ibn Yahya et décéde en 1751*, dans le man. de Leyde No.1376 (Catal. I, 164) by Dozy, suppl. II, 475.
182 Al-Muqaddasī, l.c. 6, footnote a, also *Bibl. geogr.* IV, 341. Al-Maqrīzī, I, 481.
183 *Vocabulista in arabico* (XIII cent.), publ. Schiaparelli, Florence, 1871, Dozy, Suppl. II, 475.
184 *Alf laila wa laila*, Habicht-Fleischer (ed.), V, 273.
185 Quatremère, l.c. CXXXII.
186 Abd al-Latīf, *Historiae Aegypti compendium*, 146.
187 Concerning other meanings of *warrā q* such as stationer, copier, bookseller, bookbinder, see Ibn Khaldūn, *Proleg.*, Bulaq edition, I, 352f. Quatremère, *Histoire des Mogols* etc., CXXXIII.
188 Wattenbach, 120, in 1407.
189 Hamburg, 1845, 116, 171.
190 *Kitāb al-aqālīm*, de Goeje (ed.), 262.
191 Al-Istakhrī, *Kitāb al-aqālim*, l.c. 261, 265.
192 Ersch and Gruber's *Allgemeine Encyclopädie*, l.c. 90f.
193 Abū al-Hasan, l.c. 26.
194 Wiesner, S. 180, 224ff., 227f., 229, 254.
195 Dozy, Suppl. II, 669. Vullers, l.c. II, 1315. Low, *Aramäische Pflanzennamen*, 157.
196 Al-Muqaddasī, l.c. 100.
197 Low, *Aramaische Pflanzennamen*, 290f.
198 Al-Muqaddasī, l.c. 100.
199 Al-Nabrāwī, *Nihayat al-rutba*, Ms. in the Vienna Hofbbliothek, Cod. 8131, (N.F.272), fol. 27r, 28v.
200 Ibn al-Furāt, *Ta'rī kh al-duwal wa-al-mulūk*, Arabic ms. in Vienna Hofbibliothek, A.F. 117, IX, Vol., fol. 42v.
201 Keferstein in the *Allgemeine Encyclopädie*, Ersch and Gruber, 85.
202 *Tarīkh-i Vassāf*, l.c. 434.
203 *Kitāb al-fihrist*, l.c.159
204 *Kitāb maqāmāt*, de Sacy (ed.), 552.
205 Hāfiz-Abru (d. 1431) in *Collection scientifique de l'Institut des langes orientales*, III. Persian ms. Rosen, St. Peterburg, 1886, 107.
206 See my description in the *Catalog der historischen Austellung der Stadt Wien*, 1883, 3rd edition, S.228, Nr 697.
207 Obviously I cannot go further into the study of manuscripts at this stage. I will have to save this aspect of my work for another publication.
208 *Dīwān al-inshā*, l.c. CXXXIII
209 Al-'Asqalānī, *Durar al-kāmina*, etc., ms. in Vienna Hofbibliothek, Cod, 1172, II, fol. 9v.
210 Cod. Arab. 406 of the Hof-and Staats Bibliothek, fol. 97v.
211 Rashīd al-Dīn, *Histoire des Mogols*, Quatremere, l.c., CXXXIII.
212 Khalīl al-Zāhirī, Cod. Arab. 695, fol. 4r of the National Library, Paris, by Quatremère, l.c. CXXXIII.
213 *Dīwān al-inshā*, l.c.
214 Abū al-Mahāsin, Cod. Arab. Paris, 663, fol. 49. Quatremère, l.c. CXXXIII.
215 Al-Maqrīzī, *Kitāb al-suluk*, Cod. Arab. Paris, No. 672. I, 562 by Quatremère, l.c. CXXXIII.
216 Cod. Arab. Paris, Nr. 690, Fol. 8r by Quatremère, L.C. CXXXIII.
217 Al-Asqalānī, *al-Durar al-Kāmina*, ms. in Hofbibliothek Vienna, 245, fol 3r.

218 Ms. Arab. Paris, Nr. 1573, fol.109v in *Histoire des Sultan Mamlouks* by al-Maqrīzī, transl. Quatremere, II, ii, 221.

219 *Tarīkh-i Bayhaqī*, publ. Calcutta, 1861, 168.

220 Rashīd al-dīn, *Histoire des Mongols*, l.c. CXLV.

221 Hammer-Purgstall, *Īlkhān*, l.c. I 216f.

222 P. de Alacal, *rezma de papel*, Dozy-Engelman, Gloss, 333.

223 Wattenbach, l.c.121.

224 *Dīwān al-inshā*, l.c. CXXXV.

225 *Al-Khāfaji, Shifā 'al-eghalīl*, 128H edition, 180. The word can also mean pale yellow referring not only to the colour of paper, which surely varied, but also during the period of its use, to *qartās*, a sheet or roll of paper. The technical term for papyrus of this type was *manqūf* meaning buff or pale yellow. For example, Rainer 6954 (8th century) 'two thirds of a papyrus roll of the pale yellow sort, price ½ dirham'. The comparison of 'Syrian *qartās*' to 'a camel's cheek' refers however to parchment – *Mu'allaqa* by Tarafa, edited by Arnold, 46, verse 31.

226 Quatremère, l.c. I, 2, 176; II, 2, 221.

227 Ibn Khaldūn, *Prolégomènes*, Quatremere (ed.), I, ii, 347. The recipe for ink given by Ibn al-Bawwāb is interesting: Pine soot is mixed with wine vinegar or the juice of unripe grapes, and to this is added a small amount of red ochre mixed with camphor and orpiment.

228 A certain Danieli was a master of this craft in Baghdād in the year 319H (931AH): 'He made the paper look old and wrote on it in a hand that imitated the old script.' Ibn al-Athīr, *Chron.*, Tornberg (ed.), VIII, 169.

229 Dozy and de Goeje (eds), 192.

230 A yellow and scarlet cloth also existed.

231 *Najm al-Dīn ibn Isrā'il*, al-Maqrīzī, I, 322.

232 The word *raqq* can only be interpreted as parchment, in al-Maqrīzī (*History of the Mahometan Dynasties of Spain*, II, 141). The term is explained by Zamakhsharī, l.c. 50, in Persian, as 'a sheet of gazelle skin'. Baidāwī (ed. Fleischer, II, 288) remarks about *raqq manshūr*, 'parchment that is spread out', in the Quran, Sura 52, verse 3. *Raqq* is the parchment used for writing; in the abstract, it is the material on which the divine word is written. As a final example, the *Fihrist* (l.c.21) differentiates between *raqq*, parchment and *waraq*, paper.

233 Ibn al-'Idharī, *al-Bayān al-mughrib*, Dozy (ed.), 229, 231.

234 Ibn al-'Idharī calculates the date correctly (p. 231) showing that the many dates that have been passed on to us to be incorrect. Romanus started his co-rulership in 948. Coins minted during his reign bear the same symbols as the gold bull described by Arab historians: the bust of the emperor and his co-ruler with the Greek cross between them; on the reverse is a frontal bust of Christ.

235 Bar Ali, *Lex*, Hoffman (ed.), I, 87, Nr. 2451. Payne-Smith, *Thes. s.v. baltin*, Bar Bahlul. Faustus of Byzantinum, *Geschichte Armeniens*, from the trans. by M. Lauer. Koln, 1879. Hamza of Isfahān, *Annals*, Gottwaldt (ed.), I, 48ff. This same historian, writing in the second half of the 10th century, uses the expression 'sky-coloured' for the colour hyacinth-purple (p. 50).

236 Kitāb al-'uyūn, de Goeje (ed.), 292, and Gloss 33. Dozy, Suppl. I, 588.

237 Rukn al-Dīn Baybars (d. 1328), *al-Tuhfa al-mulukiyya fī al-dawla al-turkiyya*, ms. in Vienna Hofbibliothek, Cod. 904, Mxt.665, fol. 64v 67v.

238 Ibn al-Khatīb, *Rawd al-akhyār*, ms. in Vienna Hofbibliothek, N.F.63, fol.121r.

239 *Dīwān-i Hāfiz*, Rosenweig (ed.), I, 834.

240 Therefore the Persian 'festive dress' is used as a synonym for 'red dress'.

241 Ibn al-Khatīb, l.c. fol.121r.

242 Similar to Cod. CCXXXVII in the Royal Library, Copenhagen, assembled from red paper in 738H (1337AH).

243 Mırkhwand, *Hıst. Selj.*, Vullers (ed.), Persian text, 130.

244 Dīwān-i Hāfiz (d. 1389), Rosenzweig-Schwanau, I, 336.

245 Al-Balādhurī, *Kitāb futuh al-buldān*, de Goeje (ed.), 464f.

246 Jāmī, *Yūsuf und Zulaikha*, Rosenzweig (ed.), 202, 224.

247 *Dīwān al-inshā*, l.c. CXXXIV.

248 Al-Maqrīzī, I, 91.

249 Al-Tha'ālibī, *Kitāb latā'if al-ma'ārif*, 15. There are remnants of such registers on papyrus, in book form, from Egyptian offices in the Rainer Collection.

250 *Dīwān al-inshā*, l.c. CXXXVII.

251 *Dīwān al-inshā*, l.c. CXXXIV to CXXXVII.

252 Al-Maqrīzī, II, 211. Al-Suyūti, *Husn al-muhādara*, II, 220–26.

Translator's notes

(i) Indictions, or Roman tax censuses, were held every fifteen years. This gave the cycle of indications followed in the Roman empire before the adoption of the Christian era.

(ii) Arab papermaking is descibed in *Umdat al-kuttāb wa uddat dhawī al-albāb* by Tamin ibn al-Mu'izz ibn Bādīs (1007–1061 AD). Ibn Badīs was a royal patron of the arts at al-Mansuriyya near Qairouan. He was writing about papermaking in the Maghrib; techniques might have varied elsewhere, for example, methods of paper polishing. Martin Levey, *Medieval Arabic Bookbinding and its Relation to Early Chemistry and Pharmacology* (The American Philosophical Society, Philadelphia, 1962).

(iii) The name of the river is usually given as Thalas or Talas.

(iv) 'Those papers from the C3 to the C18 found in Sinkiang consist of, besides mulberry bark, chiefly raw and fabricated fibres of hemp, flax and China grass.' Tsien Tsuen-Hsuin, 'Chemistry and Chemical Technology' in J. Needham, *Science and Civilisation in China*, Vol. 5, part 1 (CUP 1985). 'While the first paper of China was probably fabricated from disintegrated cloth . . .' Dard Hunter, *Papermaking* (London 1947).

(v) 'Another industrial product manufactured in *matbakh* of impressive dimensions was paper. Abū Sa'īd, a Maghribi visitor to Egypt, remarks that the papermills were confined to Fustāt and not found in Cairo, the seat of the caliph' (p. 81). 'It has been suggested that paper, like sugar, was manufactured in factory-like, larger establishments rather than smaller workshops . . .' (p. 112). S. D. Goitein, *A Mediterranean Society*, Vol. 1 (University of California, 1968). '. . . and in the medieval period the centers of the industry were located in Fayyum, Fustat and Fuwwa.' Terence Walz, 'The Paper Trade of Egypt and the Sudan in the Eighteenth and Nineteenth Century' in M. W. Daly, *Modernisation in the Sudan* (New York, 1985).

(vi) Karabacek names the town Manbīj when he writes it in Arabic but transliterates it as Mambīj, the form he uses throughout the essay.

(vii) Wiesner's criteria for distinguishing between linen and hemp have been found to be inadequate. Much work has been done on fibre recognition by Collings and Milner. For details of hemp and linen see *The Paper Conservator*, Vol. 3 (1978), 4 (1979). The positive identification of the two fibres in paper can, however, still present problems.

(viii) If such an easily identifiable fibre as paper mulberry has not been found in paper, Karabacek's statement is quite definitely disputable.

(ix) Arab paper moulds were usually made of split reeds or grasses, sometimes with a cloth covering.

(x) There are many Mamluk Qur'ans of the highest quality on laid paper. Karabacek examined papers mainly from Egypt and from before the 14th century. Paper technology underwent various changes in the 14th century. It might have been expected that he would have found some examples of paper with double or triple groups of chain lines among the later papers. Many papers appear to be wove at first glance but laid lines can be seen if the paper has been skinned or water stained. Some papers will occasionally show just a few centimetres of chain line.

(xi) I have frequently found 5 laid line/cm in papers from the 11–13th centuries.

(xii) Possibly Karabacek misinterpreted the reference by al-Muqaddasī about sheets of paper being stuck together. The double-sided papers that I have seen have a thickness similar to thin card (e.g. 0.4mm) and were made between the 10th and 12th centuries. His suggestion that most Oriental papers were made by sticking a wove sheet to a laid sheet is not correct. Arab papers frequently delaminate with age or use but this is not caused by failure of the paste payer. The linen or hemp fibres were seldom well fibrillated. The result was a loosely felted paper with poor internal cohesion. Both sides were then given a hard surface by paste washing. It is to be expected that this paper would split easily in two, the inner surfaces having a 'rough, woolly and felt-like appearance'.

(xiii) Rib shadows, like chain lines, are always across the width of the sheet and so provide a useful means of telling how the paper has been folded, even if it is wove paper.

(xiv) The ex-Royal Asiatic Society (now Khalili) copy of Jāmi'al-Tawārīkh has a page size of 435 × 300mm. Trimmed drawings allow us to assess the amount of paper removed in rebinding over the centuries. By this means, I estimate that the original page size was approximately 500 × 360mm giving a bifolio sheet 500 × 720mm. The rib shadows are vertical in this manuscript and these like chain lines lie across the width of a sheet of paper (see previous note).

Folio and octavo sheets show vertical shadows and quarto show horizontal. By this information we must assume that the 'large Baghdād format' ordered by Rashīd al-Dīn was approximately 500 × 720mm, or what Karabacek calls a half sheet.

(xv) If rolls of paper were made by sticking widths together, the chain lines would run across the roll. The Taylor-Schechter Collection from the Cairo Geniza has 13 manuscripts on laid, joined paper dating from Fatimid and Ayyubid times (Vols 28, 32). Six have vertical chain lines and seven have horizontal.

(xvi) There is no reason to believe that a cobalt blue was used at this period.

(xvii) Karabacek estimates the size of Complete *Tūmār* to four places of decimals!

Index